T0235679

Borderlands Resilience

This book offers new insights into the current, highly complex border transitions taking place at the EU internal and external border areas, as well as globally. It focuses on new frontiers and intersections between borders, borderlands and resilience, developing new understandings of resilience through the prism of borders. The book provides new perspectives into how different groups of people and communities experience, adapt and resist the transitions and uncertainties of border closures and securitization in their everyday and professional lives. The book also provides new methodological guidelines for the study of borders and multi-sited bordering and resilience processes.

The book bridges border studies and social scientific resilience research in new and innovative ways. It will be of interest to students and scholars in geography, political studies, international relations, security studies and anthropology.

Dorte Jagetic Andersen, Associate Professor, Department of Political Science and Public Management, University of Southern Denmark

Eeva-Kaisa Prokkola, Professor in Human Geography, University of Oulu, Finland

Border Regions Series
Series Editor: Doris Wastl-Walter, *University of Bern, Switzerland*

In recent years, borders have taken on an immense significance. Throughout the world they have shifted, been constructed and dismantled, and become physical barriers between socio-political ideologies. They may separate societies with very different cultures, histories, national identities or economic power, or divide people of the same ethnic or cultural identity.

As manifestations of some of the world's key political, economic, societal and cultural issues, borders and border regions have received much academic attention over the past decade. This valuable series publishes high quality research monographs and edited comparative volumes that deal with all aspects of border regions, both empirically and theoretically. It will appeal to scholars interested in border regions and geopolitical issues across the whole range of social sciences.

For more information about this series, please visit: www.routledge.com/Border-Regions-Series/book-series/ASHSER-1224

Borderlands Resilience

Transitions, Adaptation and
Resistance at Borders

Edited by
Dorte Jagetic Andersen and
Eeva-Kaisa Prokkola

Routledge
Taylor & Francis Group

LONDON AND NEW YORK

First published 2022
by Routledge
2 Park Square, Milton Park, Abingdon, Oxon OX14 4RN

and by Routledge
605 Third Avenue, New York, NY 10158

Routledge is an imprint of the Taylor & Francis Group, an informa business

© 2022 selection and editorial matter, Dorte Jagetic Andersen and Eeva-Kaisa Prokkola; individual chapters, the contributors.

The right of Dorte Jagetic Andersen and Eeva-Kaisa Prokkola to be identified as the authors of the editorial material, and of the authors for their individual chapters, has been asserted in accordance with sections 77 and 78 of the Copyright, Designs and Patents Act 1988.

All rights reserved. No part of this book may be reprinted or reproduced or utilised in any form or by any electronic, mechanical, or other means, now known or hereafter invented, including photocopying and recording, or in any information storage or retrieval system, without permission in writing from the publishers.

Trademark notice: Product or corporate names may be trademarks or registered trademarks, and are used only for identification and explanation without intent to infringe.

British Library Cataloguing-in-Publication Data
A catalogue record for this book is available from the British Library

Library of Congress Cataloging-in-Publication Data
A catalog record has been requested for this book

ISBN: 9780367674274 (hbk)
ISBN: 9780367674281 (pbk)
ISBN: 9781003131328 (ebk)

DOI: 10.4324/9781003131328

Typeset in Galliard
by Apex CoVantage, LLC

Contents

Figures and tables

Figures

Tables

Contributors

Dorte Jagetic Andersen is Associate Professor at the Centre for Border Region Studies, Department of Political Science, University of Southern Denmark with a background in European ethnology, European continental philosophy and political science. Her main research interest is in identity-formation in areas influenced by geopolitically drawn borders, and she has published widely on these issues in internationally recognized journals.

Péter Balogh is lecturer at ELTE Eötvös Loránd University, Institute of Geography and Earth Sciences, Department of Social and Economic Geography, Budapest, Hungary. He is also a research fellow at the Centre for Economic and Regional Studies, Institute for Regional Studies, Pécs, Hungary. His main research interest revolves around borders, geopolitical narratives and spatial identity politics.

Paul Blanchemanche is a PhD student at the Information and Communication Sciences Department of the University of Lorraine in France, the CREM. His research topics include identity studies, discourse and representational analysis, digital media and the transformation of sexuality in the digital era.

Heriberto Cairo is Professor in Political Sciences in the Universidad Complutense de Madrid, where he has been Dean of the Faculty of Political Science and Sociology. His research is in the field of political geography, with special emphasis on the study of the geopolitics of war and peace, political identities, territorial ideologies and borders. He has been a visiting professor at several universities and has published numerous books and articles.

Md Azmeary Ferdoush is a postdoctoral researcher in the Geography Research Unit at the University of Oulu, Finland. He specializes in political geography with a focus on the study of the state, borders and (non)citizenship, and currently he is working on a project which explores regional changes and transformation processes in the Arctic with an emphasis on Arctic Finland and its adjacent regions.

Steen Bo Frandsen is Professor and Head of the Centre for Border Region Studies at the University of Southern Denmark. He is a trained historian, and his main research areas are European borders and regional history of the 19th and 20th century, with a specialization in Danish-German relations, where he has published extensively.

Olga Hannonen is a postdoctoral researcher at the Karelian Institute, University of Eastern Finland. She has published articles on the themes of cross-border tourism, second homes, lifestyle mobility and migration with a special focus on the Finnish-Russian border areas.

Mariano García de las Heras is a PhD candidate and researcher in the Faculty of Political Science and Sociology of the Universidad Complutense de Madrid. He works on borders, border crossing and resistance in the Spanish-Portuguese context between 1939 and 1975.

Katharina Koch is Post-Doctoral Associate in the Canadian Northern Corridor Research Program at the University of Calgary, Canada. She completed her PhD degree in geography in 2018 at the University of Oulu in Finland. Her current research focuses on a variety of northern Canadian policy issues related to transportation infrastructure, Arctic security and geopolitics as well as the digital divide.

Jussi P. Laine is Associate Professor of Multidisciplinary Border Studies at the Karelian Institute of the University of Eastern Finland, also holding the title of Docent of Human Geography at the University of Oulu, Finland. He is a human geographer, yet in his approach to borders he combines influences from international relations and geopolitics, political sociology, history, anthropology and psychology. Within border studies, he seeks to explore the multi-scalar production of borders and bring a critical perspective to bear on the relationship between state, territory, citizenship and identity construction and has published widely in the field. He recently served as the President of the Association for Borderlands Studies.

Christian Lamour is a researcher in urban, cultural and border studies at the Luxembourg Institute of Socio-Economic Research (LISER), research associate at the Information and Communication Sciences Department of the University of Lorraine in France, the CREM and a full member of the Observatory of Regional Politics at the University of Lausanne in Switzerland. His research topics include spatial policies, governance strategies, cultural practices and bordering processes in European metropolitan regions, areas where he has published extensively.

María Lois is Lecturer in Political Geography in the Faculty of Political Science and Sociology of the Universidad Complutense de Madrid. She has published extensively on borders, heritage and critical geopolitics.

Eeva-Kaisa Prokkola is Professor of Geography at the University of Oulu and holds the Title of Docent in Human Geography and Border Studies at the University of Eastern Finland. She has extensive experience in conceptual and empirical studies of borders, border regions and identity and mobility issues. She has published widely on European internal and external border regions in internationally acclaimed journals and book volumes, and she has served as an expert in several international panels and journals, providing reviewer service for over 30 scientific journals.

Juha Ridanpää is a senior research fellow in human geography and a docent of geography, working at the Geography Research Unit, University of Oulu, Finland. His research interests include linguistic minorities, popular geopolitics, narrative regions, post-colonial studies, ethnic minorities, northern art, geographical studies of humor and literary geography. Ridanpää has published research in several international, peer-reviewed journals.

Sara Svensson is senior researcher at Halmstad University, Sweden, in the Department of Public Administration. She is a is political scientist with a specialization of public policy. Her research has focuses on the territorial cooperation and governance of cross-border regions in Europe. She has published extensively on these issues in international journals.

Preface

When we started planning this volume, we had no idea how topical the question of sudden border transitions and borderlands resilience would become in 2020 because of the COVID-19 pandemic. Over the last two years, we have followed from a distance as well as experienced in person how border controls and travel regulations have been re-established, even within what many Europeans have come to know as a "borderless" European space. Also in this context, strict border controls became a normalized state response to the pandemic in their attempts to hinder the spread of the virus.

Reinforced border restrictions and control may indeed change much of our conceptions of borders in Europe as both resources and barriers. From the perspective of borderlands, however, it is obvious that these processes resonate in historical experiences and that "borderless Europe" is but a fragment of such experiences. We only need to make a very quick historical overview to remind ourselves that many borderlands and people living in them have borders as an integral part of their DNA, with violent and otherwise turbulent histories, and regional and cultural characteristics of borderlands seem to be enduring and resilient also during periods of intense EU integration.

Moreover, borders and borderlands are often located at the margins of state maps, and simultaneously they are at the core of political and societal life. Due to the central position of borders in state politics and geopolitical power plays, border regulations often function as a state measure through which different international and domestic pressures are responded to. Borderlands and their peoples, whose lives are intimately entangled with border dynamics, therefore need to cope with and maintain resilience to stressful border events and orders that are imposed by the state governments. Concrete and silent ways of resisting border control rearticulate, sometimes easily, sometimes not, in the borderlands where resilience in the face of state rule seems ongoing and persistent, yet at the same time as changeable in time and place. Hence, borders and borderlands are without doubt both interesting and topical, from a historical perspective as well as at present.

The book's editors have a long experience of researching borders and borderlands. They became interested in the field of border studies in the course of the 2000s when the role of the state borders seemed to be diminishing in Europe and the idea of moving towards "borderless Europe" was almost taken for granted. To both editors, this idealist and almost mythical understand of borders did not seem sufficient, considering how national imaginaries and institutional settings told different narratives all over Europe. Even at its peak during the 2000s, "borderless Europe" was also confronted with the increasing securitization internationally influencing its external borders tremendously, thus questioning the sustainability of the borderless idea itself. The COVID-19 pandemic highlights the question of the resilience of borders and borderlands, thus sparking an increasing interest in studying them. Inevitably, the COVID-19 pandemic therefore also slightly changed the initial idea and focus in some chapters. Having followed the state and borderland dynamics for a long time, many authors found it necessary to also relate their analysis to the new situation at the border.

The editors would like to thank all authors for their contributions and collaboration. The editors and authors come from different disciplinary backgrounds, yet the ensuring interest and fascination with borders and borderlands connect our work. The well-institutionalized research field of border region studies provides a common starting point to our research and thinking, which is further developed in this volume to widen an understanding of borderlands and their resilience processes in different historical and geopolitical contexts.

The book's ensemble of contributions presents insights inviting us to rethink resilience in the light of social and cultural processes characterizing borderlands and what it takes for borderland people to have "a good life." Although much focus in this volume is on the European border regions, the insights the volume build on stretches beyond Eurocentric understandings of borders, and our work on borderland resilience has also been inspired by the writings on the borders and borderlands located in African and Asian contexts, as well as more broad outlooks on global interconnections. As collective, the contributions illuminate the importance of rethinking resilience, not only as possibility, but as necessity because geographical borders as well as bordering practices are essential to societal change at a variety of scales and in a variety of historical contexts. We hereby open for and invite further discussions about multiple and diverging forms of resilience, in research as well as in policymaking, with the hope that the following discussions each and together contribute to the steadily growing literature on borderlands, resilience and both.

Sønderborg, Denmark and Oulu, Finland, 27 May 2021
Dorte Jagetic Andersen and Eeva-Kaisa Prokkola

1 Introduction
Embedding borderlands resilience

*Dorte Jagetic Andersen and
Eeva-Kaisa Prokkola*

Borderlands under cumulative stress

Borderlands almost worldwide are currently experiencing considerable transitions because of the intensifying global trend of tightening control at state borders. The slight optimism of moving into a "borderless" world, present especially in Europe in the 1990s and early 2000, has vanished, and the question of border security and national protectionism have veered to the front of political agendas. Border scholars have documented a shift from an open borders policy and transnational collaboration towards the building of new border fences (Brown 2010; Jones 2012; Bromley-Davenport et al. 2019; Paasi et al. 2019).

In a global context, border regions have suffered from the financial crisis of 2008 and faced the large influxes of refugees and migrants. In Europe, many countries that for decades strived to abolish the barriers that borders create, have since the 2015 "long summer of migration" (Scheel 2015) continuously reinforced and strengthened border controls, also at internal Schengen borders. Moreover, the United Kingdom's Brexit referendum and Trumpist politics underlined how strongly purported the need seemed to be to restrict and control human mobilities. The geopolitical tensions after the Russian annexation of the Crimean Peninsula have greatly influenced cross-border mobilities, and trade relations between Russia and its neighboring states were particularly affected (Koch and Vainikka 2019; Raudaskoski and Laine 2018). In 2020, we have experienced a dramatic change in the discourse of borders, that is, the closing of borders globally as a response to the COVID-19 pandemics. Alongside these "fast stress events," borderlands face different slow crisis like climate and environmental change and prolonged industrial restructuring. These are often more difficult to measure because of their cumulative nature.

To respond to what is understood as a continuous, and as indicated previously, cumulative crisis (Hudson 2010), academics from various disciplines, including governments and international organizations, have turned their theoretical and practical interests towards resilience. Resilience is abundantly examined in the context of climate change, risk management, urban transitions and regional economic crisis, among others. This has opened promising avenues both in research and in policymaking, but also critique. The increasing interest towards resilience is

DOI: 10.4324/9781003131328-1

often explained in terms of neoliberal governance; that is, policymakers are seeking to support the resilience of communities and are indicating a need to increase self-reliance and sustainability at the community level (OECD 2014; Wagner and Anholt 2016). Moreover, resilience is sometimes explained, in somewhat problematic ways, to stem from the fact that "the opening of state borders has made 'places and regions more permeable to the effects of what were previously thought to be external processes'" (Christopherson et al. 2010, p. 3; Prokkola 2021, p. 4). Consequently, the resilience debate contributes to the naturalization of borders as security measure. However, if resilience is pretheoretically reduced into a neoliberal strategy, we will lack an understanding of complex mundane practices of people and communities (Wandji 2019). People can act on and transform the world, and they are not just objects of governance but also capable of resisting and finding alternative ways of living (Chandler and Reid 2016). Hence, the study of people and institutions in place and how reinventions are established in shifting conditions may offer a different version of "what counts as resilience" than what the neoliberal discourse suggests.

The mushrooming resilience research touches upon various ecological, economic, political and social questions. Regardless of its multidisciplinary and global reach, the questions of borders and their impact on resilience have only recently started to gain attention (Wandji 2019; Prokkola 2019; Korhonen et al. 2021). Recent research points out that resilience research and interventions usually take bounded regions and communities as a point of departure (Healy and Bristow 2019; Bristow and Healy 2020), even when the problems to be addressed would be highly transnational and global in their scope. It is important to recognize, however, that resilience is not inherently bound with some bordered, administrative entities. Cooperation across scales and times forms an essential factor for human adaptability as connections and networks across borders facilitate social interaction, the flow of ideas and resources (Davoudi et al. 2013; Korhonen et al. 2021). The neglect of the topic of borders is surprising considering that the significance of inter-scalar relations and border crossing was acknowledged already in the influential early writings of socio-ecological resilience in terms of "panarchy" (Walker et al. 2004). To understand contemporary resilience processes, it is crucial to recognize the significance of borders and border crossing, as well as the political contestations, cultural values and norms that are manifested in and through bordering. Border studies has potential to engage with and contribute to the social scientific resilience discussion by examining what roles borders have in the social and regional resilience processes and how borderland communities are adapting, renewing and resisting the border transitions and changing border environments. The view from borderlands complicates the prevailing, territorially bounded understandings of social resilience.

The collection provides new knowledge about the present, highly complex border transitions that are taking place at the EU internal and external border areas and globally, and their influence on the lives and mobilities of borderland people. The enduring question is, how different groups of people whose lives are always-already entangled with borders and border crossings maintain wellbeing and adaptive capacities in the face of border transitions, including reinforced securitization as well as new openings. A related question concerns the possibilities for resisting transition in the borderlands. This volume brings together border cases through which it examines the impacts that multilayered and multi-sited borders have in order to rethink our understandings of resilience. The focus is on different borderlands and groups of borderland people, and their practices, conceptions, memorizations and visual and textual representations of border transition, adaptation, coping strategies and capacities and resistance. Thus "seeing like a border" (Rumford 2012), the book aims at opening a new path to understanding borderlands resilience.

Multidisciplinary resilience

The idea of resilience has its background in psychology, mathematics and ecology. There is no consensus about the exact definition of resilience, however, and different disciplines underline different understandings and epistemic standpoints. Although the term "social resilience" – referring to the ability of groups and communities to cope with external stresses as a result of social, political and environmental change (Adger 2000; Walker et al. 2004) – is prolifically employed, the application of resilience theories in social scientific research is considered rather problematic from methodological and normative perspectives. Accordingly, there is a voluminous literature discussing in what ways social scientific resilience research is meaningful and important, or not (Joseph 2018; Brown 2013). The reason for the concern is understandable: whereas the ecological resilience research focuses on environmental hazards, in the social sciences discussions of resilience have been extended to include different human-made crisis like economic shocks, terrorism and even migration "crisis," hence placing resilience first and foremost in political, cultural and normative context (Joseph 2018, pp. 13–14).

The traditional social-ecological perspective would direct its focus on border landscapes, infrastructure and environmental design (Grichting and Michele Zebich-Knos 2017) but pays less attention to the complex political and sociocultural nature of borders. From a socio-ecological context, we appropriate the idea that social resilience expresses and is connected to materiality above and beyond in how borderland communities (attempt to) restore the status quo. Inherent protective and wellbeing factors are emphasized, leading to questions of what are the assets, resources and strengths possessed by the society or community making it able to cope when faced with adversity, including assets such as social and political capital, or certain cultural, institutional or economic factors. Different historical times are also compounded of different social and political assets. Accordingly, resilience is best understood as a continuous process to sustain wellbeing, not a trait or an outcome (Southwick et al. 2014, p. 4).

The socio-ecological resilience thinking has been developed further in regional studies and planning, where it has opened a field of researching regional vulnerabilities and strengths from a more holistic perspective. The focus has been on factors that could possibly explain why other regions and communities succeed in adapting to chancing economic and institutional conditions while others do not (Hassink 2010). Resilience is often approached as the capacity of complex regional "systems" to accommodate shocks and to move back to the conditions before a shock. Recently many scholars have turned their interest towards the long-term capacity of regions and cities to renew their institutional structures and to find new growth paths (Boschma 2015). More understanding, however, is needed on the role of extra-national public policy, interregional and cross-border cooperation in regional resilience processes. Moreover, in borderlands and other transnational contexts, the fundamental question of resilience, namely "resilience for whom," is highlighted and intimately intertwined with the institutional and everyday politics of solidarity (Bristow and Healy 2020).

There is also an interdisciplinary interest towards "structural resilience," emphasizing how communities or societies have incorporated resilience over time, thereby making resilience an inherent characteristic of the community. Prasad et al. (2009) for instance refer to "the resilient society," which has the:

> capacity of a community or society to adapt when exposed to hazard. . . . A resilient society can withstand shocks and rebuild itself when necessary. Resilience in social systems has the added human capacity to anticipate and plan for the future. (p. 32)

The claim is that one can speak about "resilient cultures," thereby linking the notion of resilience to identity-formation. We discuss the links between resilient cultures and borderlands at length later.

In addition, resilience remains a complex construct because of the origins of the resilience debate in developmental psychology. In this context, resilience concerns the individual, and focus is on the individual's ability to adapt or manage sources of stress and trauma and thus "bounce back" over a lifespan. Methodologically, development psychology follows people (also located in groups, like families or institutions) over time. Psychological resilience discourses have attempted to "turn the tables," so to speak, and move away from a purely deficit-based model of mental health to instead focus on strengths and competence-building when people experience stress. In this context, the link to meaning-making is emphasized; that is, the way human beings are making sense of things in the face of chaos, conflict, violence, etc. Resilience is understood to be supported by an "ability to hang on to a sense of hope that gives meaning and order to suffering in life and help articulate a coherent narrative to link the future to the past and present" (Southwick et al. 2014, p. 10). The focus on meaning-making, narratives and imaginaries informs our accounts of resilience in this volume, also when our concern is with society, culture and community. Hence, and as we explore further later, we find the situational approach, the analyzes of resilience processes in their historical, political and cultural contexts, useful (Cote and Nightingale 2012) in the study of borderlands.

Borderlands resilience

To conceptualize *borderlands resilience*, we have scrutinized both the concept of social resilience and the various meanings of "borderland." The term borderland is usually used to refer to a region that straddles along or across one or more international borders. In the modern system of states, borderlands have been often seen as peripheral and, in many ways, vulnerable regions (Sahlins 1989). This understanding has been partly challenged since the establishment of the free trade areas (EU, NAFTA) and border openings that shifted border regions from a peripheral position to a more economically and politically favorable one (Hanson 2001). Researchers have documented how the EU internal border regions and many global border cities have become important nodes for international flows (Sohn 2014). Moreover, some scholars like Gloria Anzaldúa (1987) have employed the concept of borderland as a synonym to cultural diversity and hybridity, thus paying attention to a very specific cultural and psychological meaning of borders and border crossing in the lives of people. In comparison, border studies usually underline the materiality of borders, yet no longer understood to materialize merely at the physical borderlines but extend as social-political institutions. Thus, the enacted effects of a border can take place in multiple locations, often in rather surprising ways (Andersen et al. 2012; Amilhat Szary and Giraut 2015; Laine 2016). To outline, a borderland refers to terrain and landscape where borders materialize in very powerful ways in the lives of people.

Borderlands are without doubt specific ecological, political and social environments where people and communities have historically had to find their ways to cope with and within the structures of two or more states, as well as their transitions. Changes in border areas, such as infrastructural ones and those related to cross-border mobilities and attitudes influence borderlanders' lives and anticipate futures in different ways. It is rarely asked, however, how political borders and strengthened border securitization hinder the vernacular and regional resilience strategies in the face of environmental, economic and socio-cultural change. When border communities and mobile people need to cope with manmade material border infrastructures, renewal and resistance may emerge as a response to such border transitions. Considering a recent trend away from open borders scenarios towards border securitization and unsolved political conflicts, these questions become highly pertinent. Secondly, borderlands, like other places and regions, are vulnerable to environmental hazards, economic shocks and social conflicts that are by no mean territorially fixed. Yet the means and strategies available for states and communities to cope with changes and risks are often territorially confined, focusing efforts on the geographical areas where countries have their sovereignty. From the perspective of borderlands resilience, the border location and the material and discursive nature of the border may have considerable, still largely unidentified impacts. The situational approach is important: European internal border areas and African postcolonial borderlands (Wandji 2019; Laine et al. 2020), for example, have rather different historical and cultural narratives of borders, yet in all borderlands people have strived to maintain their meaningful social-material relations, everyday activities, cultural heritage and identity. Borderlands are often seen as hybrid cultural spaces, yet connections across borders are not something that would automatically sustain in all border contexts when the conditions change. For understanding borderlands resilience from the perspective of identity and everyday lives, we need to pay attention to the different material and social institutions that both connect and bound and separate people.

"People's resilience"

When studying resilience in a context of borderlands, we need methodological approaches that are sensitive to agency and simultaneously pay attention to the context and institutional structures that support the processes associated with resilience. Unlike other life forms, human beings can make and plan interventions into the processes to diminish, sustain and enhance resilience (Davoudi 2012). Therefore, to understand resilience processes, attention need to be paid to the capacity and interest of different groups of people to make and discuss, to forecast and anticipate vulnerabilities and consciously change their behavior and/or location. As Bristow and Healy (2020; see also Davoudi et al. 2013) explain, resilience has a strong behavioral element; that is, resilience emerges not only from inherent institutional and structural conditions – like the establishment or withdrawal of border regulations – but also from the stimulus of people who are able to impact the trajectory of change. People may establish and find new creative ways to cope with the environmental, social and economic change and border transitions. As well as understanding how the borderland people adapt to stress situations, it is important to pay attention to how they interpret, articulate and make sense of different shocks, and how this influences their responses.

The study of borderlands resilience therefore includes the examination of the situational practices, experiences and narratives of borders, border crossing and belongings from the perspective of adaptation and resistance among different groups and communities. Of interest are, first, practices, social relations and belongings in which borders and border crossings make a difference and, second, the ways through which people reorient their practices and relations with regards to border transitions. It is possible to gain understanding of the entanglement of resilience processes with the long-lasting socio-cultural and geopolitical power relations and contestations by analyzing how these relations are manifested in border experiences and narratives providing guidance to adaptive pathways and resistance. At political borders, different versions of resilience may intertwine in surprising ways, and people may intentionally ignore and resists the political discourses of resilience in their everyday environments, place-making, social and cultural practices. Hence, our choice has been to problematize how resilience expresses in various ways in borderlands, including policymaking processes and the everyday life of people living there. We believe the "bottom-up approach" is needed because the discussion of borders remains under-developed in the resilience literature. As Levine and Mosel tells us: "People are resilient to the degree that they avoid falling into unacceptable living conditions" (2014), but the understanding of what are "unacceptable living conditions" varies from place to place, from threat to threat and from stress to stress.

To understand borderlands resilience, close attention should be paid to people's many ways of "making sense" of what is "a life worth living," and problematizing resilience at this level involves posing simple questions about what it takes for people and communities to achieve a "good-enough life" in a real-life context. As Panter-Brink puts it: "I think that the most important and effective way to approach resilience is to start with listening to what people have to say about their everyday lives" (Southwick et al. 2014, p. 10). Resilience is a situated, interpretative process, involving what is already had (resources, skills, etc.) in combination with what is known (mobilizing possibilities). Hence, the attempt is to understand resilience intrinsically as dependent on the circumstances of those affected and the context in which stress is experienced, rather than normative ideals about "the good life." People's stories of their cultural goals lead us to matters that elucidate their resilience (Ibid., p. 10).

Moreover, and following Wandji (2019), we emphasize the situated meaning of, not just resilience, but also of threat, shock, stress and disruption and how resilience also involves questioning notions of what is threatening and what is causing stress. Wandji (2019, p. 289) points out that while the conventional definition of resilience causes the notion of threat to be limited to an idea of catastrophe or shock, threats and shock can actually be slow moving and barely perceptible to people with an outside perspective. Instead, threat is reconfigured in terms of a "plurality of disruptions," something that embraces the idea of the need to be resilient in the face of constant challenges, but questions the idea that this has to be seen in terms of a dramatic external event. Zooming in on resilience processes in everyday life, we avoid getting trapped in the normative zero-sum game of applying a universalized notion of "the threat" communities are exposed to, making the threat purely external and turning crisis into processes local communities are unable to control or be partly responsible for. Any universalized notion of threat runs the risk of depoliticizing the threat itself, presenting it as if it was natural and unavoidable (Wandji 2019). Moving beyond presumed universality is important in relation to our understanding of borderlands agency, as well as how geopolitical borders may themselves present a threat in the borderlands (cf. earlier).

Our main concern could thus be termed "people's resilience," that is, different social groups' ability to self-organize and mobilize skills and resources to create opportunities when faced with adversity and to act in solidarity when their community is disturbed and even disrupted. Hence, the most important insight we take with us from border studies is that borders are complex, practical constellations and never either "good" or "bad":

> That is why when we look at borders in terms of their supposed decline as barriers to movement we must balance this expectation with the evidence that many borders continue to act as gates, sometimes open, sometimes closed . . . and the business of enforcing the laws that swing such gates open and closed is in fact big business for many nations.
>
> (Donnan and Wilson 2010)

In doing this, we keep a wary eye to questions concerning the unique status of borderlands in relation to other geographical areas. Borderlands are not equal but diverse since some borderlands are highly urbanized areas while others consist of hinterlands. Most borderlands have been influenced historically by conflict and populations movement (can be both voluntary and forced), they may be more diverse population-wise than the national inland, and they thus tend to adopt different approaches than the center when it comes to politics, identity-formation and ultimately also survival strategies. Memories and the geographical imagination may also be very different from that of the center, and they may be more trained to withstand shock. Over the years, various concepts have developed to characterize such places, concepts such as "bufferzones" and "frontier regions."

Borderlands identities as resilience

Closely related to discussions of "people's resilience" are questions concerning identity-formation and the role played by identities and identifications in borderlands resilience. In the resilience literature, identity issues are mainly dealt with from two perspectives: one a developmental psychological perspective, focusing on the individual's identity-formation over time, and the other a socio-ecological perspective, where the concern is the identity of cultural systems. As the form of identity-formation we are concerned with in this volume is cultural and expressed in the self-identifications and othering processes of communities and social groups, we draw more on the socio-ecological understanding of identity and especially the notion of "cultural resilience," which "has emerged to refer to this continuity of a co-constituted set of long-term relationships between the cultural identity of a people and the set of socio-ecological relationships within which this identity was founded" (Rotarangi and Stephenson 2014). It has for instance been shown how indigenous societies in the face of transformation manage to maintain "key elements of structure and identity that preserve their distinctness" (Ibid.). Here the focus is on the integrity of the "system" and the maintenance of cultural structures and identity in the face of adversity, ultimately identifying general "change drivers" involved when these communities manage to preserve their cultures and identities, and thereby survive as distinct cultural communities.

However, when dealing with identity-formation, we challenge the static structural-functionalist understanding of collective identity-formation, located in system-thinking, where culture (including collective identity) is understood as a stabilizing part of a broader set of socio-ecological relationships. Rather, our understanding of identity-formation is inspired by post-structuralism (Hall 1996), raising question concerning representation, meaning-making narratives and memory, ultimately asking if identities developing beyond and above geopolitical power relations do not make for dynamic and complex components of both resilience and threat in everyday practice. This is to say that self-identification and othering processes are understood as resources appropriated by actors (ultimately as practices) rather than mere structural assets, thus also emphasizing agency and real-life complexity to questions concerning identity-formation. Again, the approach avoids universal claims about identity, presupposing that it is a positive asset when it works as a stabilizing factor. We illustrate how the self-identifications of people in the borderlands may be an important asset and resource involved in attempts to deal with geopolitical changes to borders and how processes of identity-formation might be understood as resources making borderlands resilient. Moreover, we also illustrate how exclusive, inflexible identity narratives may pose a threat to the integrity and wellbeing of borderlands, something that is highlighted in for instance historical struggles between national identity-formation and regional identity-formations, or in a more contemporary setting in relation to the homogenous, ethnic identifications vs. more heterogenous, cosmopolitan identifications and heritage making (Andersen and Prokkola 2021). Identity-formation is thus understood as a multiple practice and thereby much more than mere preservation and a matter of the survival of a cultural system.

This finally brings us to the complex relationship between resilience and continuity. The idea of "rebound and resume" is often promoted by resilience discourse with the person, community, company or service provider expected to carry out business as usual. Yet resilience is also about adaptation or transformation, and a tension exists between embracing change and staying true to the previous status. Bieber (2020) for instance makes a positive case for the 1999 Greek-Turkish earthquake as a critical juncture, which enhanced cross-border cooperation in the region; hence the earthquake as exogenous shock had a positive effect on social structures in the borderlands. This tension between continuity and change is evident even in the original ecology discussions (Holling 1973), and it seems inherent to any discussion of resilience, also making it essential to understanding resilience as situated rather than universality.

As noted earlier, Wandji introduces the idea of disruption, which might be said to challenge the idea of continuity or could be read as living with continuous disruption. He argues that the border community does not seek transformation and is resilient essentially in terms of adaptation as a form of continuity rather than change. The border-threat and all the obstacles this presents to communities becomes part of daily life and enables the continuity of social life across two countries divided by a political boundary and the complex, often disruptive, practices associated with this. What we are interested in is thus how such stress or disruption is experienced and dealt with by people in the local communities, including what would be the geographical imagination of people – what do people actually conceive as a threat – as well as how the same people absorb stress and disruption into everyday life and thereby work to mitigate and neutralize its impact on the geographical imagination and beyond. Ultimately this discussion brings us back to questions of identity-formation, and how identification is played out in local context. The question we raise, also by problematizing identity-formation as long-term, historical processes of relating to geopolitical borders in everyday life, is whether inclusive and flexible self-identifications do not prove to be far more durable and sustainable than any essentialist quest for identity as preservation, especially in the face of adversity; when communities are in crisis, one-sided identifications closing of border imaginaries to multiple possibilities appear to work against the tackling of crisis.

Introducing the contributions

The book's chapters approach this conceptual terrain by bringing together a range of cases, which both theorize various forms of borderlands resilience and exemplify contextually how they manifest. As argued earlier, our focus is on groups of people living in borderlands, analyzing their practices, conceptions, memorizations and representations of border transition, adaptation, coping strategies and capacities as well as resistance. The individual empirical studies provide thick and context-specific understanding of regional and social resilience and non-resilience, considering the complex effects of policy strategies, institutional structure, historical development, culture and identity (cf. Hill et al. 2008). As ensemble, the chapters provide a rich repertoire of studies on borders and borderland resilience, bridging well-established border theories and conceptualizations with the social scientific ideas of resilience. The studies thereby also provide insights into similarities between how groups of people and communities experience, adapt to or resist transitions and uncertainties of border closures and securitization in their everyday lives. Thinking different, yet related forms of borderlands resilience through each other acts as a comparative lens aiming to trace relations between spatial and timely performances of resilience constituting not just separated but also interrelated practices, materializations and affects. Hence, by bringing together different case studies of borderlands and resilience, it is possible to strengthen dialogue between different versions of social resilience concerning border areas, thereby generating new understanding through bridging governmental and bottom-up community resilience debates.

The collection begins from the conceptual horizons described earlier that resilience discussions are neglected in border literature and, at the same time, borderlands have so far been neglected in interdisciplinary social scientific resilience literature. Hence, the aim in this first part is to open the discussion of how the volume adds to the existing literature by offering theoretical and empirical insights into borderlands resilience and by examining what roles borders play in resilience. In her chapter, "Border security interventions and borderland resilience," Eeva-Kaisa Prokkola illustrates how the top-down politics of resilience interlink with current ambivalent border governance and security thinking where the imaginaries of risk in a "borderless world" are often employed to explain the increasing importance of developing resilience. Attention is paid to the processes and discourses of resilience regarding border security interventions in three different geo-historical contexts. Firstly, it considers the EU neighborhood policies and resilience building initiatives, conceptualizing these in terms of "borderwork" that stretches beyond the external borders of the Union. The chapter then shifts focus from top-down resilience discourses towards actual social resilience processes in the European borderlands. The second case examines borderland resilience vis-à-vis geopolitical events in the Finnish-Russian borderland after the

Crimean Crisis. Thirdly, the chapter examines the changing political environment and border security efforts at the Finnish-Swedish borderland. Particular attention is directed to the border security interventions during the 2015 migration influx and the COVID-19 border regulations. The chapter sheds light on the contextual nature of border interventions and borderlands resilience and highlights the importance of recognizing the politics of resilience and values in the formation of the conception of resilience.

Katharina Koch's chapter, "Cross-border resilience in higher education: Brexit and its impact on Irish–Northern Irish university cross-border cooperation," provides an example of the consequences of border transitions for the very successful cross-border cooperation between Ireland and Northern Ireland initiated with and since the Good Friday Agreement. The chapter approaches resilience from the perspective of mobility and examines cross-border cooperation between universities in Ireland and the implications of Brexit for student, faculty and staff mobility. Higher education contingency plans are understood as a preparatory form of resilience, emphasizing the role of renewal, adaptation and, to some extent also, resistance regarding relations at the shared border. Hence, the chapter illustrates the transformative power, agencies and preparations of borderland populations and professionals working in different sectors and creating new paths for cooperation and wellbeing in a changing border landscape. Moreover, the examination of cross-border resilience in higher education in the context of Brexit underlines the importance of taking a critical stance to the top-down political resilience debate. When the borderlands and the experience of actors living in them are neglected in these processes, there is a risk that instruments supposed to work towards resilience have the exact opposite effect.

María Lois, Heriberto Cairo and Mariano García de las Heras' chapter "Politics of resilience . . . politics of borders? In-mobility, in-security and Schengen 'exceptional circumstances' in the time of COVID-19 at the Spanish-Portuguese border" contributes to understanding resilience in relation to borders at different scales. It focuses on the problematic relation between the resilience of borderland people as against state bordering yet illustrating the importance of revitalizing historical trajectories in a contemporary context. In the light of current border securitization and by using the COVID-19 border closures as example, the chapter explores the role of historical memory and cross-border daily activities in borderlands resilience. In recent years, the European Union cross-border cooperation policies have turned internal borderlands into iconic places for aid and action programs, and border communities have been reframed in continuous processes of meaning-making, becoming discursive sites where various actors negotiate what is to be narrated and what spatial identities are mobilized. Currently, these narratives of open borders clash with intensifying border closures and state securitization in Europe, leaving the borderland populations to, yet again, reiterate their identities anew. The chapter brings forth how the resistance to the decisions of central governments as well as the historically formed "border tactics" and creative resistance are central to understanding borderlands resilience in the exceptional COVID-19 circumstances. Together the chapters in part one thereby set the scene for understanding the significance of geopolitical border transitions, the shift from open border policy towards strengthened regulation, and that borderlands processes of adapting to, renewing and resisting the changes and disruption of border landscape may differ considerably from the high-scale political expectancies.

Following up on the importance of "seeing like a border" (cf. earlier; Rumford 2012), the chapters in the volume's Part 2 problematize the consequences of border transition and closure from the location of resilience among a range of social groups living in borderlands. The chapters offer an understanding of resilience among various cross-border communities, including cross-border commuters, farming communities, welcome cultures, property owners, transnational movements and stateless people. Their adaption, renewal and resistance are scrutinized through local and mundane interactions, incorporating also forms of silent resistance and recognizing the agency of vulnerable people. Together, the examples provide an understanding of how border closures and transitions trigger multiple yet related social resilience practices in geopolitically and socio-culturally different borderlands, including well-integrated European Union internal border areas, areas with European Union external borders and South Asian borderlands. While exploring resilience in these different contexts, the chapters bring in new theoretical takes on the processes through which resilience is built and maintained, as well as to the limits of resilience.

In the first chapter in the section, "Resilience at Hungary's borders: Between everyday adaptations and political resistance," Péter Balogh and Sara Svensson analyze resilience processes through cross-border communities' relations in three different contexts in the Hungarian borderlands. In the first example, the focus is on cross-border agglomerations at Hungarian borders and how multi-ethnic communities adjust to geopolitically enforced changes of border security. These communities whose existence has for long been encouraged by the European Union prove to be highly vulnerable to such changes because they depend on the border's openness. The second case illustrates how food production has become a site of important nationalist symbolism towards which local cross-border food communities may imply resistance or endorsement. Thirdly, with a focus on solidarity movements that support refugees, the chapter illustrates how borderlanders and their civil society organizations responded to Hungary's policy of hardening borders from 2015 onward. The cases illustrate how borderlands resilience is a highly complex phenomenon, involving many social actors, who are affected by border transitions in various ways and thus expressive of multiple social resilience practices. Such practices do not just clash in their relation to geopolitical decision-making; they sometimes also stand in conflictual relations in the borderlands themselves, where social groups cope very differently with the stress of border closure.

Olga Hannonen's chapter, "Mobility turbulences and second-home resilience across the Finnish-Russian border," explores social resilience and its limits by demonstrating the local effects of national and international mobility regimes and the resilience practices of trans-border second-home movers in the

Russian-Finnish borderlands. Whereas the border closures and shifts experienced on the EU-internal borders are still moderate, the external borders have experienced tremendous shifts in recent years. With a brief period of openness and slight optimism about developing cross-border relations in the 1990s and early 2000s, the securitization of the Finnish-Russian border is again intensifying, considerably affecting mobile people whose social and economic lives have become entangled with border crossings. The examined case demonstrates a form of resilience incorporating silent resistance, a temporary coping strategy for second-home owners. These newly adapted practices of Russian second-home owners are reactions to changed circumstances, e.g., visa regimes and changing back policies, which are not sustainable over time but often rather inconsistent solutions to far greater problems. With the examples, the chapter provides an understanding of the limited space for coping mechanisms and agency at external EU borders confronted with intensified mobility regulations and border transitions.

In the chapter " 'Stateless' yet resilient: Resistance, disruption and movement along the border of Bangladesh and India," Md Azmeary Ferdoush argues that frameworks investigating structural change in borderlands often lose sight of the nuances of daily life, especially when it comes to conceptualizing the resilience of stateless populations. With the focus on the Bangladesh-India border, the chapter complicates the prevailing narrative of "vulnerable populations" as merely an object of governmental resilience-building interventions, showing how even stateless borderland populations who are not endowed with citizenship rights possess the capacity of acting to refuse the sovereign, as opposed to being a silent recipient of violence. It shows that expanding the focus of resilience studies to a marginal borderland population allows us to unearth overlooked, yet significant, layers of complexities and materiality on the ground. The chapter sums up this section by way of a criticism of Agamben, illustrating how paradoxes of sovereignty are reiterated as tensions in everyday life. Hence, the chapter opens up powerfully for an understanding of how some populations around the world may have an easier time dismantling, disrupting and "getting around" borders than those who are trained to believe in them as second nature, and as being of eternal value and unquestionable power.

Following up on the processes of dismantling, disrupting and getting around, the chapters in Part 3 dig deeper into borderlands resilience by zooming in on questions of identity-formation and cultural representations in border and diaspora communities evolving historically. This section underlines self-identification and othering processes as resources appropriated by actors in the borderlands, thus emphasizing agency and real-life complexity in their relation to questions concerning identity-formation. It thereby helps underline the conclusions from the previous sections that borderlands resilience is not determined by top-down politics but should be understood as practices of active socio-cultural resistance, adjustment and meddling with possibilities in intersection between geopolitics and everyday life. While doing this, the section is able to relate current debates on resilience and borders to historical processes, on the one hand, by asking how the historical processes are exemplary for borderlands resilience in a contemporary context and, on the other, by showing how borderlands resilience itself does not constitute instances of fire-fighting but rather depends on longer trajectories acting as stabilizing assets in the everyday life of the borderland populations. The chapters thus illustrate how resilience is tied with historical memory and ongoing struggles over people's identifications and this, again, in multiple ways.

In his chapter, "Schleswig: From a land-in-between to a national borderland," Steen Bo Frandsen examines resilience in a historical regional setting by way of developments in the border province of the composite Oldenburg monarchy: Schleswig, today the border region of Denmark and Germany. The chapter scrutinizes the dramatic border changes that the region experienced and how resilience formed over time in the region. It shows how state and nation-building processes, including national historiography, have done the utmost to erase regional identifications. Schleswig confronted the destructive power of national ideologies in the early 19th century, making the regional identity almost non-existent today. The Schleswigians were literally forced to choose sides in conflicts between two states, the Danish and the German, each of which were going through very different developments. The chapter shows that borderlands resilience is tied with historical memory and the difficult struggle for identifications in the borderlands. One element of resilience is a fight against the other national ideology, an element that has been strong in the nationalizing narrative of the Danish state and where the border region of Schleswig became instrumental. Another element is the resilience of the regional identifications pressured by the conquering states and nationalizing narratives leaving limited space for borderlands resilience.

Juha Ridanpää's chapter, "Borderlands, minority language revitalization and resilience thinking," also illustrates the ability of borderlands populations to mobilize identity-formation and historical memories as means of resilience, here with a focus on minority languages. The chapter focuses on Meänkieli language revitalization in the Swedish-Finnish borderland. As has been the case with several subaltern languages, a perception exists of Meänkieli speakers as passive, oppressed and harshly treated by the majority population, and feelings of shame remain embedded in the self-perception of the speakers. In such cases, the question of linguistic sovereignty is inherently connected with the bitterness directed at the denial of belonging and cultural roots, and thus with processes of colonization, marginalization and trauma. In the case of minority groups and languages, resilience thinking thereby provides an alternative to common top-down language policies; that is, it paves the way for a bottom-up approach where decisions are made by the affected populations. The chapter pays specific attention to the changing role of the Swedish-Finnish border, which, in relation to the Meänkieli language and identity represents a symbolic marker for a shameful past. The key question concerns how cultural activists involved in language revitalization re-narrativize their shifting identities by linking resilience thinking together with the practices of active socio-cultural resistance. Hence, the chapter illustrates how conflicting viewpoints over the socio-political status of minority languages in borderlands may, through creative resilience, interpret into simultaneous acceptance of uncertainty and hope for a better future for minorities.

The next chapter, Christian Lamour and Paul Blanchemanche's "A resilient *Bel Paese?* Investigating an Italian diasporic translocality in between France and Luxembourg," sheds light on resilience in the Italian diaspora found in a cross-border urban basin in-between France and Luxembourg in the course of an intensified urbanization of the border area. This resilience is approached through Appadurai's conceptualization of translocality as "scapes," mobilized to show how cross-border relations and identity are maintained in flows located beyond the spaces of governance, and by linking their country and cultures of origin and their places of everyday life. Empirically, these flows are brought forth in the article by focusing on one structuring element, which has helped renew the meaning of an Italian diasporic space over the past 40 years: Italian film festivals. Based on the analysis of the posters of the festival of Villerupt since the 1970s and its inclusion in contemporary narratives of local residents with Italian roots, the research shows that the resilience of the Italian cross-border translocality involves constant reiteration and reshuffling of connections between the inherited Italian culture and the urban space of the France/Luxembourg cross-border area. The example illustrates how resilience is a continuous process involving multiple appropriations and directions, as well as constant renegotiation even within the same social group and cultural community.

In the book's final chapter, "Line-practice as resilience strategy: The Istrian experience," which is also concluding Part 3 of the volume, Dorte Jagetic Andersen sheds light on bordering practices located in intersections between geopolitical decisions and the everyday life of people living on the Italian, Slovenian and Croatian peninsula, Istria. Inspired by Sarah Green's notions of traces and tidemarks, as well as studies in the western Balkans emphasizing the populations' ability to mimic power and play with identifications, the chapter opens towards an understanding of borderlands resilience as practice with multiple layers and possibilities ready to be adapted. On Istria, the constant redrawing of borders by different geopolitical powers open possibilities for various kinds of "line-practice," and new border closures are not just perceived as problematic for the borderlands; rather, lines are worked with, crossed and overcome, and sometimes even used strategically to articulate diversity in an otherwise integrated space. The redrawing of borders, in whatever shape they may take, tends to integrate into everyday life practices. Hence, the chapter illuminates how intersections between geopolitics and everyday life express in a landscape of multiple line-practices, where resilience towards external stress is first and foremost one of coping, meddling, disrupting and in other ways working with geopolitical transformations imposed by powers located far away from the peninsula itself, and this through a constant renegotiation of what it is to be Istrian. The chapter thereby sums up the multiplicity of resilience practice exposed in the previous chapters, and how borderlands resilience needs to be perceived in situated expressions and enactment of social relations, movement, identity-formation and historical memory in borderlands effected by border transitions and closure.

An epilogue for the collection has been written by Jussi P. Laine. He underlines how in the current world with the proliferation of the imaginaries of threat and danger, "bordering the uncontrollable" has become a common response to environmental, political and social crisis at various levels regardless of the fact that border closures usually cause more problems than solutions. He reminds us that although the resilience building could offer a more long-term solution than border security efforts, resilience is not a "universal magic bullet" for complex "intermestic" problems, many of which have a long historical presence. Laine suggests that by rethinking resilience through borderlands – their histories and people's experiences of adaptation – might enable us "to escape the clear-cut modernist *dispositifs* and dualism" between what are internal problems of the state territory and what are external.

References

Adger, N., 2000. Social and ecological resilience: are they related? *Progress in Human Geography*, 24 (3), 347–364.

Amilhat Szary, A. and Giraut, F., 2015. *Borderities and the politics of contemporary mobile borders*. New York: Palgrave Macmillan.

Andersen, D.J. and Prokkola, EK., 2021. Heritage as bordering: heritage making, ontological struggles and the politics of memory in the Croatian and Finnish borderlands. *Journal of Borderlands Studies*, 36 (3), 405–424.

Andersen, D.J., Klatt, M., and Sandberg, M., eds., 2012. *The border multiple*. Aldershot: Ashgate.

Anzaldúa, G., 1987. *Borderlands – La Frontera. The new Mestiza*. San Fransico: Aunt Lute Books.

Bieber, F., 2020. Global nationalism in times of the covid-19 pandemic. *Nationalities Papers*, 1–13. DOI: 10.1017/nps.2020.35.

Boschma, R., 2015. Evolutionary economic geography: towards an evolutionary perspective on regional resilience. *Regional Studies*, 49 (5), 733–751.

Bristow, G. and Healy, A., 2020. Supranational policy and economic shocks: the role of EU structural funds in the economic resilience of regions. *In*: G. Bristow and A. Healy, eds. *Handbook of regional economic resilience*. Cheltenham: Elgar, 280–298.

Bromley-Davenport, H., Mac Leavy, J., and Manley, D., 2019. Brexit in Sunderland: the production of difference and division in the UK referendum on European Union membership. *Environment and Planning C: Politics and Space*. 37 (5), 795–812.

Brown, K., 2013. Global environmental change I: a social turn for resilience? *Progress in Human Geography*, 38 (1), 107–117.

Brown, W., 2010. *Walled states, waning sovereignty*. Cambridge: MIT Press.

Chandler, D. and Reid, J., 2016. *The neoliberal subject: resilience, adaption, vulnerability*. London: Rowman and Littlefield.

Christopherson, S., Michie, J., and Tyler, P., 2010. Regional resilience: theoretical and empirical perspectives. *Cambridge Journal of Regions, Economy and Society*, 3 (1), 3–10.

Cote, M. and Nightingale, A., 2012. Resilience thinking meets social theory: situating social change in socio-ecological systems (SES) research. *Progress in Human Geography*, 36 (4), 475–489.

Davoudi, S., 2012. Climate risk and security: new meanings of 'the environment' in the English planning system. *European Planning Studies*, 20 (1), 49–69.

Davoudi, S., Brooks, E., and Mehmood, A., 2013. Evolutionary resilience and strategies for climate adaptation. *Planning Practice & Research*, 28 (3), 307–322.

Donnan, H. and Wilson, T., eds., 2010. *A companion to border studies*. Boston, MA: Blackwell.

Grichting, S. and Zebich-Knos, M., eds., 2017. *The social ecology of border landscapes. The anthem series on international environmental policy and agreements*. London: Anthem Press.

Hall, S. 1996. Who needs identity? *In*: S. Hall and P. du Gay, eds. *Question of cultural identity*. London: Sage.

Hanson, G., 2001. U.S-Mexico integration and regional economies. *Journal of Urban Economics*, 50 (2), 259–287.

Hassink, R., 2010. Regional resilience: a promising concept to explain differences in regional economic adaptability? *Cambridge Journal of Regions Economy and Society*, 3 (1), 45–58.

Healy, A. and Bristow, G., 2019. Borderlines: economic resilience on the European Union's eastern periphery. *In*: G. Rouet and G. Pascariu, eds. *Resilience and the EU's eastern neighbourhood countries*. Cham: Palgrave Macmillan.

Hill, E.W., Wial, H., and Wolman, H., 2008. Exploring regional resilience. IURD Working Paper Series, Paper WP-2008-04. Institute of Urban & Regional Development.

Holling, C.S., 1973. Resilience and stability of ecological systems. *Annual Review of Ecology and Systematics*, 4, 1–23.

Hudson, R., 2010. Resilient regions in an uncertain world? *Cambridge Journal of Regions, Economy and Society*, 3 (1), 11–25.

Jones, R., 2012. *Border walls: security and the war on terror in the United States, India, and Israel*. London: Zed Books.

Joseph, N., 2018. *Varieties of resilience: studies in governmentality*. Cambridge: Cambridge University Press.

Koch, K. and Vainikka, V., 2019. The geopolitical production of trust discourses in Finland: perspectives from the Finnish-Russian border. *Journal of Borderlands Studies*, 34 (5), 807–827.

Korhonen, J., Koskivaara, A., Makkonen, T., Yakusheva, N., and Malkamäki, A., 2021. Resilient cross-border regional innovation systems for sustainability? A systematic review of drivers and constraints. *Innovation: The European Journal of Social Science Research*, 34 (2), 202–221.

Laine, J., 2016. The multiscalar production of borders. *Geopolitics*, 21 (3), 465–482.

Laine, J., Moyo, I., and Nshimbi, C., 2020. Borders as sites of encounter and contestation. *In*: C. Changwe, I. Moyo and J. Laine, eds. *Borders, sociocultural encounters and contestations: Southern African experiences in global view*. London: Routledge, 3–14.

Levine, S. and Mosel, I., 2014. Supporting resilience in difficult places. A critical look at applying the resilience concept in places where crisis is the norm. HPG commissioned. Available from: www.odi.org/sites/odi.org.uk/files/odi-assets/publications-opinion-files/8881.pdf (accessed 24 August 2020).

OECD, 2014. *Guidelines for resilience systems analysis*. OECD Publishing. Available from: https://www.oecd.org/dac/Resilience%20Systems%20Analysis%20FINAL.pdf (accessed 16 September 2021).

Paasi, A., Prokkola, E-K., Saarinen, J., and Zimmerbauer, K., eds., 2019. *Borderless worlds for whom? Ethics, moralities and mobilities*. London: Routledge.

Prasad, N., Ranghieri, F., Shah, F., Trohanis, Z., Kessler, E., and Sinha, R., 2009. *Climate resilient cities: a primer on reducing vulnerability to disasters.* Washington, DC: World Bank.

Prokkola, E-K., 2019. Border-regional resilience in EU internal and external border areas in Finland. *European Planning Studies,* 27 (8), 1587–1606.

Prokkola, E-K., 2021. Borders and resilience: asylum seeker reception at the securitized Finnish-Swedish border. *Environment and Planning C: Politics and Space.* April 2021, DOI: 10.1177/23996544211000062.

Raudaskoski, M. and Laine, J., 2018. Changing perceptions of the Finnish-Russian border in the post-cold war context. *In:* J. Laine, I. Liikanen and J.V. Scott, eds. *Post-cold war borders: reframing political space in the EU's Eastern Europe.* London: Routledge, 129–146.

Rotarangi, S.J. and Stephenson, J., 2014. Resilience pivots: stability and identity in a social-ecological-cultural system. *Ecology and Society,* 19 (1), art. 28.

Rumford, C., 2012. Towards a multiperspectival study of borders. *Geopolitics,* 17, 887–902.

Sahlins, P., 1989. *Boundaries: the making of France and Spain in the Pyrenees.* Berkeley, CA: University of Califormia Press.

Scheel, S., 2015. Das Konzept der Autonomie der Migration überdenken? – Yes Please! [Rethinking the concept of autonomy of migration? – Yes Please!]. *Movements,* 1 (2), 1–15.

Sohn, C., 2014. Modelling cross-border integration: the role of borders as resource. *Geopolitics,* 19 (3), 587–608.

Southwick, S.M., Bonanno, G.A., Masten, A.S., Panter-Brink, C., and Yehuda, R. 2014. Resilience definitions, theory and challenges: interdisciplinary perspectives. *European Journal of Psychotraumatology,* 5 (1).

Wagner, W. and Anholt, R., 2016. Resilience as the EU global strategy's new leitmotif: pragmatic, problematic or promising? *Contemporary Security Policy,* 37 (3), 414–430.

Walker, B., Holling, C., Carpenter, S. and Kinzig, A., 2004. Resilience, adaptability and transformability in social – ecological systems. *Ecology and Society,* 9 (2).

Wandji, G., 2019. Rethinking the time and space of resilience beyond the West: an example of the post-colonial border. *Resilience: International Policies, Practices and Discourses,* 7 (3), 288–303.

Part 1

Borders and resilience in "exceptional circumstances"

2 Border security interventions and borderland resilience

Eeva-Kaisa Prokkola

Introduction

During recent decades, border regions have witnessed various stress events, stemming from complex and often unpredictable geopolitical events, and economic, social and environmental disruptions and shocks. The global COVID-19 pandemic and national efforts to mitigate the spread of the virus by introducing extremely strict border controls provides a topical and unique example of tightened border control. Yet since the beginning of the millennium, border scholars have documented how states have introduced tightened border control and protectionism as a response to various kinds of threats and malaises (Ackleson 2005; Amoore 2006). A border closure often creates a stress situation and increases and multiplies the experience of disruption in the lives of borderland people. The present chapter takes a critical stance towards border security inventions and underlines the importance of cooperation and open borders from the perspective of borderland resilience.

The chapter starts from an understanding that the relationship between political borders and processes of resilience is highly complex and multidimensional. Like resilience, borders and borderlands are also subject to multiple definitions, depending on the scale and on whether the focus is on material border infrastructure or social communities and institutions. Borders are basic political institutions (Anderson 1996) that materialize in a multilayered manner through legal, administrative, economic, social and cultural practices (Paasi et al. 2019). Many scholars have documented how, after a border institution is established, a border gradually becomes an inseparable part of the activities and mindscapes of people (Sahlins 1989).

Historical developments in borderlands reveal how borders have a pervasive influence in shaping the organization of human life and expressions of identity. From the historical perspective, the state border *in itself* can be seen to entail a "plurality of disruptions" that the border inhabitants, authorities and economies need to cope with and negotiate in the organization of their activities and settings (Wandji 2019). Borders and consequential disruptions are not merely located at the physical borderlines, however, but are enacted and materialized in contextual and multilayered ways across various spheres of life (Newman and Paasi 1998; Andersen and Sandberg 2012; Paasi et al. 2019). Exposure to sudden changes does not usually imply immediate acceptance, adaptation and management of

DOI: 10.4324/9781003131328-3

change, however, but can rather be approached as a continuous process involving different phases where social power relations and the decisions of actors play a major part (Cote and Nightingale 2012; Bristow and Healy 2020). When the notion of resilience is scrutinized in relation to multidimensional borders, attention therefore needs to be paid simultaneously to the politics of resilience, thus asking "resilience for whom, what, when, where, and why" (see Meerow and Newell 2019; Cutter 2016).

The chapter scrutinizes the processes and discourses of resilience regarding border security interventions in three different geo-historical contexts. Firstly, it considers the EU neighborhood policies in which resilience thinking has become the new leitmotif (Wagner and Anholt 2016). The EU-funded resilience-building initiatives in its eastern and southern neighborhoods are conceptualized in terms of a "border-work" that stretches beyond the external borders of the Union (see Bialasiewicz 2012). The discussion of the EU's neighborhood highlights the politics of resilience and the "resilience for whom" question (Bristow and Healy 2020). The chapter then shifts focus from top-down resilience discourses towards actual social and regional resilience processes. The second case examines borderland resilience vis-à-vis geopolitical events in the EU external Finnish-Russian borderland, where the 2014 Crimean Crisis and consequent international sanctions halted cross-border mobilities and travel from Russia to Finland. Thirdly, the chapter examines the changing political environment and border securitization efforts at the EU internal Finnish-Swedish borderland. Particular attention is directed to the border security interventions during the 2015 "long summer of migration" (Scheel 2015) and the COVID-19 border regulations. The three examples illustrate the contextual nature of border interventions, cross-border regulations and resilience, their entanglement with the politics of scale as well as how different values shape problem and solution narratives in a state of continuous change.

The different geopolitical border cases attest to the spatial, temporal and contextual multiplicity of resilience (cf. Simon and Randalls 2016). They provide insight into how geopolitical environment and border transitions influence the lives of borderland people and into how the politics of resilience is entangled with border securitization. This is a topical question in Europe and globally because of the changing political landscape and recently tightened border controls. Cross-border regions and borderlands are often regarded as "laboratories" of European social and territorial development (van Houtum 2000), emphasizing solidarity, conflicts/conflict resolution and the process of integration. Following this understanding, the resilience processes of cross-border regions and connections may also be indicative of the resilience of European space making and identity more generally (see Jensen and Richardson 2004).

Resilience thinking and the "world of permeable borders"

The increasing interest towards resilience theories can be attributed to a generalized experience of uncertainty and continuous crisis (Hassink 2010) as well as to the processes of globalization that "have made places and regions more permeable to the effects of what were once thought to be external processes" (Christopherson et al. 2010, p. 3). In governmental documents and scholarly debates alike, the essential role of state borders is often implicitly represented as a key explanation for the increasing interest towards the resilience approach. The imaginary of the risks of a shrinking, "borderless" world is employed to make sense of and explain the need for fostering resilience. The United Nations Global Sustainability (2012) report "Resilient People, Resilient Planet" depicts the enhancement of resilience as a response to global problems within an increasingly complex and interconnected world, for example. Resilience debates tend to naturalize political borders as physical lines that divide different "systems of resilience" and protect the "inside" from the disturbances coming from the "outside." The narrative of resilience in a world of permeable borders reveals that the articulation of resilience is not sensitive to the full spectrum of political interests and multiplicities of values; instead it often cements the state-centric view of resilience. Hence, state security becomes the "completing value" shaping " 'problem' and 'solution' narratives during times of disruption" (cf. Rogers et al. 2020).

The perception of external risks and the shrinking of the world both refer to the taken-for-granted assumption of the state territorial order, and thus borders. Borders and border transitions are the product of human societies, however, and thus are incompatible with the logic of unpredictability and non-human forces like environmental disasters. Recent stress events like COVID-19 have pointed out, however, that the construction of borders and border regulations often plays a crucial role in the state-centric processes of adaptation, resistance and renewal. Border security interventions and border closures are argued to provide a solution to various kinds of problems and malaises, many of which have resulted from domestic policies. In the articulation of border security interventions, the political nature of resilience becomes highlighted, provoking ethical questions regarding whose resilience and entitlement to wellbeing are made visible and in what ways.

Contrary to the narrative that presents borders as a solution to domestic and global disturbances, a considerable amount of research originating from various disciplines suggests open borders and diversity increase societal resilience. Simin Davoudi et al. (2013), for instance, note that cooperation across scales is important for human adaptability, since connections across borders facilitate social interaction and innovation. This notion is supported by the original socio-ecological system resilience theory and its notion of panarchy (Walker et al. 2004); that is, the resilience of a people, a community or a region are understood to depend on their dynamic organization and structuring "within and across scales of space and time" (Allen et al. 2014, p. 578). Regardless of this fact, the mainstream resilience research is often embedded in a territorial conception of space where resilience is measured in relation to some territorially bounded administrative entity or community (Bristow and Healy 2020).

Modern state borders are Janus-faced in character, "poised between openness and closure" (van Houtum et al. 2005, p. 12). In border studies literature, open borders are usually considered a resource for regional and socially more harmonious development. Open borders represent a resource because the border location entails proximity to foreign markets and labor, the possibility to take advantage of cost differentials, the diffusion and stimulation of new knowledge and ideas as well as new regional identities and brands (Sohn 2014; Prokkola and Lois 2016). Borders are human organizations and thus coordinated by socially and politically constructed rules that to a great extent explain how and to what extent some places and communities resist, adapt and renew when confronted with stress situations. Somewhat paradoxically, open borders and cross-border connections provide a resource that has proved to be especially valuable in times of national "border crisis" (Prokkola 2019). Accordingly, local perspectives in border areas challenge the prevailing, territorially bounded understandings of resilience and attests to the complex role of borders in resilience.

Borders are not merely manifested in the form of "hard" political lines but usually entail differences in terms of "soft" borders, that is, in the culture and system of values that play a pivotal role in the processes of resilience (Rogers et al. 2020). Historical developments regarding the border are crucial for understanding the expressions and articulations of resilience in borderlands (cf. O'Dowd 2010), together with the notion that different cultures and institutions coexist and sometimes collide at borders. Borderlands are fruitful sites for studying in what ways geopolitics, high-level policies and security interventions influence local resilience processes and capacities. They are sites where the state powers manifest in concrete ways and where people often have rather different attitudes towards "the other side" and cross-border mobilities compared with the national centers (Anzaldua 1987; see also Andersen 2022 and Frandsen 2022). Any study of borders and resilience therefore needs to be sensitive to social and cultural values and political contestation. The situational and contextual examination of resilience can reveal the possible mismatches between resilience policies and local resources as well as contradictory values guiding resilience processes. It also raises the question of in what ways cross-border connections and processes are vulnerable to various economic, socio-cultural and environmental disruptions that the establishment, maintenance or securitization of a border creates.

EU neighborhood and the politics of resilience

The notion of the EU neighborhood provides a fitting example of how top-down resilience discourses often represent borders and border management as a response to various insecurities. Resilience appears as a foreign policy goal in the revised European Neighborhood Policy of 2015 with the aim to "strengthen the resilience of the EU's partners in the face of external pressures and their ability to make their own sovereign choices" (European Commission and HR/VP 2015, p. 4). It is argued that resilience and resilience building has become the new leitmotif of EU neighborhood policies (Wagner and Anholt 2016). The state and social resilience in the EU's "east and south" is also one of the priority areas of the EU Global Strategy, launched in 2016 (EU 2016).

A review of the EU strategies and policy documents on the neighborhood suggests that resilience-building initiatives are closely connected to the EU's externalized border and migration management. The revised policy underlines the EU's interdependence with its neighbors, thus explaining how "growing numbers of refugees are arriving at the European Union's borders hoping to find a safer future" (European Commission and HR/VP 2015). It directs attention to the root causes of migration, and describes resilience as an effort to prevent and manage migration in the long term.

> The new ENP (European Neighborhood Policy) will make a determined effort to support economies and improve prospects for the local population. The policy should help make partner countries places where people want to build their future, and help tackle uncontrolled movement of people.
> (European Commission and HR/VP 2015, p. 4)

Borders and bordering are bound with geopolitical power relations and materialize through the practices of exclusions and inclusions. The EU's promotion and funding of resilience building in the neighborhood countries and their societies can be understood in terms of a bordering or "border-work" that stretches beyond the external borders of the Union (see Bialasiewicz 2012). Extending the analysis of resilience to the EU's external border and "border-work" provides new understanding regarding the role of geopolitics, extra-national public policy, interregional cooperation and solidarity (Bristow and Healy 2020).

Examination of the EU's Global Strategy points out that the EU resilience discourse has a strong security connection. The European border-work that has taken place in the Union's southern and eastern neighborhoods is part of an attempt to secure the internal by securing the external. Resilience building stands as a preventive and stabilizing action that is contingent upon the availability of knowledge concerning potential security "problems" in the neighborhood. In this respect, the discussion of regional resilience in border areas, and threats to that resilience, often turns into a question of border and migration management (cf. Bourbeau 2015) in a world of permeable borders. Resilience policies function as soft power to prevent migration by supporting adaptation *in situ* instead of developing legal migration routes from North Africa and the Middle East to the area of the European Union. Anholt and Sinatti (2020, p. 311) even argue that "for the EU, resilience-building is primarily a refugee containment strategy that could jeopardize the stability of refugee-hosting states." EU resilience governance invites the neighboring states to co-border its external borders and support their migrant populations so that the migrants would stay where they are and not seek protection and wellbeing by attempting to enter the EU area. As Biscop (2017) puts it:

> If Europe's neighbours are resilient to certain threats, those threats will not reach Europe itself. In more standard geopolitical jargon, a resilient neighbour would be called a buffer state. That is a role that may appeal to certain

governments, if the EU offers a high enough price. The EU has clearly begun to use Turkey as a buffer state in the field of migration, for example, paying it a hefty sum in return.

(Biscop 2017)

The EU resilience discourse introduces "novelty, adaptation, unpredictability, transformation, vulnerability and systems" into a new governmental vocabulary that makes the governance of uncertainty a fundamental rationale (Welsh 2014, p. 16). The neighborhood concept provides a fitting example of resilience governance that aims to shift responsibility to the regional and community levels, as if they were self-sufficient entities – something that is considered problematic from a normative policy perspective (Wagner and Anholt 2016, pp. 415–416).

Resilience can be approached from the point of view of information generation and legitimacy, where the key question is whose resilience is concerned and what objectives may be included in the presentation of concerns (Cote and Nightingale 2012, p. 482). Scholars argue that "investing in the resilience of states and societies beyond the Union's borders is a way forward to enable societies to minimize the impact of crises and thus deter potential threats from the EU" (Eickhoff and Stollenwerk 2018). This comes out especially in the debates on climate migration and the EU-driven resilience building initiatives abroad, which support adaptation in place and intra-African mobility. The financial support for the building of a more resilient neighborhood is by no means altruistic; instead it is expected to "pay itself back" by increasing European internal security and by preventing large-scale migration to Europe.

The EU neighborhood policy and the management of the EU's external borders highlight the fundamental question of resilience for whom – a question that is intimately intertwined with the institutional and everyday politics of solidarity (Bristow and Healy 2020). The Union's resilience discourse does not say much about the actual social and community resilience of the societies and communities in the EU neighborhood. Instead, it reveals that the notion of resilience for whom and why is an essential factor to consider when studying resilience and the discourses of resilience regarding different borders and borderlands.

Schengen borderlands and resilience

The geopolitical environment and historical processes of borders greatly influence the mechanisms of resilience. The opened/closed nature of the border varies according to geopolitical environment, partly determining the development trajectories of a borderland and its resources to cope with various environmental, political, economic and social changes and stress situations (Prokkola 2019). The typology of borderland interaction, developed by Oscar Martinez (1994, see also Timothy 1999), provides a useful starting point for contemplating how the border type can influence borderland resilience. Firstly, "alienated" border regions are usually characterized by geopolitical tensions. Border crossing is restricted or does not exist at all. The North and South Koreas offer an example of an alienated border region. Secondly, "coexistent" border areas usually have neutral relations that enable some sort of interaction, yet both countries have adopted inward-looking strategies in their problem-solving. From the resilience point of view, this means that if a border community faces a sudden stress event and disruption, they must mainly rely on local capacities and domestic institutions. Thirdly, "interdependent" regions have cross-border relations that are characterized by networking and cooperation. Here border communities have established initiatives to solve common problems through legislative cooperation; thus they might be better prepared and resilient when facing crisis situations. Fourthly, "integrated" border areas where all border restrictions have been removed and the regions are functionally merged. Here the border communities have established multi-scale connections across the border and maintain stable and well-institutionalized cooperation. The geopolitically steady EU internal borders are characterized by this kind of collaboration and functional regionalization, yet its borders still materialize though their legal, political, economic and cultural layers (Paasi and Prokkola 2008; Svensson 2015).

Geopolitical borders and their economic and social trajectories are rarely stable, however. Furthermore, the volume and temporality of cross-border interaction often varies over time. The historical developments of borders show that border openings and the development of cross-border cooperation are often gradual and relatively long processes, whereas border security interventions and border closures can take place hastily as a response to geopolitical events or experienced global and/or national insecurity. Accordingly, to gain understanding of the processes of resilience requires more than simply considering borderland resilience in terms of the border typologies and classifications: attention also needs to be given to border dynamics and transition. Also, the transition from one border typology to another "stage of interaction" both requires resilience and impacts the resilience of a borderland in the long term. For example, integrated border communities can possess capacities and resourcefulness that can be used to anticipate and respond to sudden environmental hazards or slow-onset events.

The Finnish-Russian border, the EU's longest external border, provides a fitting example of a security border where economic and socio-cultural development have been influenced by sudden geopolitical events. The Crimean Crisis in 2014, the consequent economic sanctions and the decline of the Russian ruble immensely influenced the socio-economic conditions of the Finnish-Russian borderland, where Russian cross-border shoppers have been the driving force of economic growth (Hannonen 2022; Koch 2018). Between 2013 and 2016, Finnish-Russian cross-border traffic decreased by 32.2 percent (Finnish Border Guard 2020), and Russian visitor arrivals in all accommodation establishments in Finland decreased by 56.4 percent (Statistics Finland 2020). The concentration and specialization on Russian trade and tourism in the border area increased the vulnerability of the borderland economy to geopolitical turbulence.

In response to the immobility shock, the tourism industry was expected to realign and find new growth paths and realignment strategies that could increase the resilience of the border region's economy (Prokkola 2019). Regardless of the local renewal strategies, the development of cross-border connections in the EU external borderlands is to a great extent dependent on wider-scale political decision-making and geopolitics. Local coping mechanisms and opportunity structures are limited and closely interlinked and dependent on wider EU-level and national policies towards Russia. The experience of living within a geopolitically sensitive border engenders a specific borderland culture and structures of meaning making. It is suggested, for example, that people living next to geopolitically sensitive borders are somewhat socialized to a specific border mentality and practical approach towards border-related high geopolitics. In the Finnish-Russian borderlands, older generations of Finns often mention "the lessons learned from history" as a way of explaining how local people are successfully coping with geopolitically sensitive situations (Prokkola 2019). Simultaneously, historically formed relationships and knowledge within and across borders are crucial factors that enable regional actors to find solutions even in difficult conditions (Boschma 2015). Altogether, examination of the EU's external borderlands shows that geopolitical environment and cross-border connections influence the processes of adaptation, resistance and renewal (see also Bristow and Healy 2020), that is, the resilience of borderland communities. While some coping mechanisms and paths of adaptation and renewal are available to borderland people and authorities, many are not. This is partly explained by the type of borderland interaction and the historical dynamics of the border. Moreover, the selection of the coping mechanism is steered by the "completing values" (Rogers et al. 2020), and thus different geopolitical actors may prefer different approaches.

The resilience of European cross-border regions

European territory contains more than 100 institutionalized border and cross-border regions that have acquired membership in the Association of European Border Regions (AEBR, founded in 1971). The multiplicity of cross-border regions and their connectedness can be seen to reflect European aims of integration, tolerance and solidarity. Institutionalized cross-border regions such as the Euroregions serve as fruitful sites for gaining understanding of the resilience processes of cross-border regions and their inter-scalar connections.

Since the "long summer of migration" of 2015 in Europe, there has been a shift from developing cross-border cooperation and cross-border regions towards tightened border security measures. Many EU countries that for decades strived to abolish the barriers that borders create have recently reintroduced border controls. Border surveillance and checks have again become part of the mundane experiences of mobile people travelling and commuting across the borders. Presently, Euroregions are experiencing a sudden stress from the closing of borders as a response to the COVID-19 pandemic.

The Finnish-Swedish Tornio Valley cross-border region serves as an example of an EU internal cross-border region and Euroregion where border checks and border closures have been reinstituted. It can be defined as a northern Euroregion and an integrated border area (see Martinez 1994) where interaction and cooperation are well institutionalized. The Tornio Valley is often seen as a historically borderless region with traditional communities straddling the border. The border has been relatively open since World War II, and it is often referred to by locals as the most peaceful border in the world. Local people and communities have historically had to find their own ways to cope with the different institutional structures of Finland and Sweden, as well as the different iterations of the border. Administratively, the cross-border region has its roots in the 1987 establishment of the Council of Tornio Valley, which encompassed the border municipalities along the Finnish-Swedish border. Finland and Sweden joined the EU in 1995 and the Schengen agreement in 2001 together with Norway. Since then, the regional institutions and connections have been strengthened, notably within the framework of the EU Interreg North Program (cf. Baldersheim and Ståhlberg 1999).

Several decades of open border policy was called into question in autumn 2015 when hundreds of asylum-seekers started to arrive daily at the northern Finnish-Swedish border crossing point. The situation was considered exceptional in Finland because historically Finland has not been a destination country for migrants. Altogether, Finland received a tenfold number of asylum applications compared with previous years (~3,000 →32,476 applications), with most of the asylum-seekers travelling across the northern Swedish-Finnish land border-crossing point in Tornio. In response to the increased numbers of asylum-seekers, the Finnish government relocated hundreds of security sector officials to Tornio to control the border crossing and the asylum-reception process. The border was securitized by the state authorities – a move that was instigated in part by nationalist right-wing movements that urged a total closure of the border. Extreme nationalists mobilized threat imaginaries, and their presence in Tornio created an extra burden for actors responsible for the reception, maintenance and care practices in the asylum reception process (Prokkola 2020).

The open border represented an important resilience factor in the securitized environment of asylum reception. The daily reception work was organized in collaboration with state and local authorities and non-governmental actors such as the Finnish Red Cross and local volunteers. The volunteers were able to work on both sides of the borderline to support the reception activities. Accordingly, in the organization of the reception, cooperation and existing networks within and across the border gained a crucial role. The actors gained topical and trustworthy information from contact persons in Sweden, for example, regarding the time, place and volumes of the new arrivals. Timely information was considered crucial for planning the reception and work schedules in Tornio. Many local actors underlined that the peaceful history and trade relations in the border region provided a resource that enabled them to better cope with the changed and stressful situation (Prokkola 2020). It was paradoxical that at the same time as many citizens and politicians were presenting the closure of the border as a national security means, regional- and local-level cross-border cooperation was essential from the point of view of the everyday security of civil society and smooth reception work. The 2015 Tornio case shows that cross-border cooperation and a culture of cooperation contribute to the ability of regions and communities to better cope with changing situations. Paradoxically, the border security measures put in place may even weaken the sense of security in the long run because they ultimately hinder cross-border collaboration and the maintenance of trust relations that support resilience in the borderlands and in wider society.

Today, the COVID-19 pandemic has caused an exceptional stress situation locally and globally. In open border areas like the EU internal Finnish-Swedish borderland, however, the border closure has had considerable impacts on the everyday routines and economic and social environment of the borderland people, causing additional stress and confusion. The closing of the internal EU Finnish-Swedish border on March 19, 2020 created not only a barrier but also a new and unfamiliar environment of different regulations and logics of filtering and controlling cross-border mobilities. The border traffic decreased approximately 90 percent (Finnish Border Guard 2020). The border closure and travel restrictions due to COVID-19 are by no means exceptional; they are part of a normalized response to the global COVID-19 pandemic, an attempt to hinder the spread of the virus. Yet, the Finnish-Swedish border is a highly interesting case because Finland and Sweden initially adopted highly different coping strategies and degrees of border control. In Sweden, the government and experts first relied more on the willingness of citizens to govern themselves, whereas in Finland the government followed many other countries and immediately imposed strict regulations and border control.

Compared with the "long summer of migration," COVID-19 is a very different kind of disruption at the Finnish-Swedish border because the border was practically closed for the local people for the first time since WWII. Whereas during the 2015 border security intervention, the mobility of Finnish and Swedish citizens was not regulated, during the pandemic all people needed to have a legitimate, usually work-related reason to cross the border. The mobility of health care commuters from Finland to Sweden formed a highly critical regional question. If Finland had restricted the mobility of health care professionals, the health care sector of the Swedish Tornio Valley region would have collapsed. The COVID-19 crisis has clearly underscored that the reinstitution of borders and border closures *in itself* represents a disruption from the perspective of borderland people and their everyday lives and the regional economy. The control and regulation implemented at the Finnish-Swedish border meant that borderland people needed to continuously negotiate the different national strategies and oftentimes ambiguous instructions and statements from the Finnish government. Local and national media reported about divided families and the experienced difficulties and stresses of local people. After a month, it was reported in the news that many local people started to feel that the border closure was more disturbing and traumatic than the disease itself (Passoja 2020).

In May 2020, Finnish citizens realized that since Sweden had not established border controls, they could simply cross the border as their constitutional right (Juntti 2020). The situation was rather confusing, and many Swedes experienced it as unfair. Some anticipated that the Nordic solidarity and good relations were at serious risk and that it will take a long time to rebuild the trust. In May, the Finnish-Swedish border crossing point become a curious place of state control: a hybrid space where people did not know exactly what the situation was and where the rules of border crossing were negotiated case by case.[1] The regional authorities and actors needed to actively lobby the central state, which initially failed to recognize the unique borderland culture and connections across the border. Finally, in August 2020, the Finnish authorities established a new borderland citizenship status, "member of a border community" (Finnish Border Guard 2020). The new rule meant that people who live in the Finnish and Swedish border municipalities are legally permitted to cross the Finnish-Swedish border even during the pandemic. A similar decision was made regarding the Finnish-Norwegian border.

This study of the Finnish-Swedish border during "crisis" events shows that in cross-border areas the mechanisms of adaptation and renewal are linked with both state-centered institutions and cross-border networks and institutions at multiple scales. It remains open, however, to what extent the asymmetric COVID-19 border control and restrictions and consequent border securitization will impact the idea of the "borderless" border and historically formed trust relations in the Tornio Valley border region, like other Euroregions. Connections and trust relations across the border are not something that would automatically withstand changes in geopolitical conditions. Cross-border regions are established and continuously maintained through institutional practices and everyday social relations; thus their institutional organization, established role, and territorial and symbolic shape may change (Paasi 2003). Indeed, there is a concern that long-lasting border barriers could impugn the identity and function of the cross-border regions and the European project per se (Opiłowska 2021).

Discussion: contextual borderland resilience

Countries commonly introduce heightened border control as a response to various challenges and crises that are explained to have external origins. The responses are in many ways paradoxical because, as border scholars and others have long underlined, global phenomena like environmental hazards, economic instabilities or pandemics rarely respect state borders. Contrary to the popular understanding that draws a connection between strict border control and state internal security, open borders, transnational cooperation and solidarity usually increase societal resilience and wellbeing in the long term.

The border approach provides new understanding regarding the significance of geopolitical environment, political contestation and values – something that has been neglected in studies of resilience (Bristow and Healy 2020; Phelan et al. 2013). Any analysis of European borderland resilience needs to recognize the geopolitical site, governmental intervention, and acts of bordering at the points of resilience's articulation (cf. Simon and Randalls 2016). This chapter has examined border security interventions in three different European border and border policy contexts, with the focus on local and regional resilience construction and the politics of resilience. Resilience processes have been scrutinized in relation to geopolitical conditions, border typology, regional histories, and regional and socio-cultural connections. The examples illustrate how resilience construction gains different political and normative meanings in different border contexts.

The difference between the EU internal and external border regions underscores the difference that geopolitical environment makes from the viewpoints of resilience. Examination of the EU external Finnish-Russian border shows how the geopolitical tension between the EU and Russia, after the Russian annexation of the Crimean Peninsula, arrested ongoing and planned EU–co-funded development programs and froze regional cross-border trade and cooperation. The regional stakeholders in the Finnish-Russian borderland showed adaptation to the prevailing situation and hoped that the political situation would change for the better. Their agency was limited but not totally withdrawn; they continued cross-border cooperation by keeping contact with trusted local partners on the Russian side, for example. In comparison, at the Finnish-Swedish border, cross-border cooperation has often intensified during "crisis" events to alleviate the political tension. Established long-standing cross-border networks and relations were considered highly important from the perspective of the 2015 asylum reception. Similarly, local cross-border connections and lobbying have proved valuable during the COVID-19 pandemic. The specific conditions of the borderland were ultimately recognized by the Finnish government, resulting in the introduction of a new citizenship category, "member of a border community." Members of the border community on both side of the border now possess the right to cross the border even during the pandemic. This suggests that the borderland's resilience has its own logic that is interconnected with yet simultaneously different from the national and European Union political agendas (see also Lois, Cairo and de las Heras 2022).

The EU resilience policy, the narratives regarding the long summer of migration and COVID-19 border restrictions all highlight the importance of recognizing the politics of resilience and values in the formation of the conception of resilience. Comparison of the EU internal and external border areas in Finland fittingly illustrates how different state borders can be subject to different geopolitical regimes, institutions, cultural values and trust relations, which all influence borderland resilience. Cultural knowhow, strong trust relations and institutionalized connections across borders form an important resilience asset that is more likely present in open border contexts and cannot be generalized to all borders. Borderland resilience is also developed against state institutions and core-periphery relations. For example, the trustworthiness of Finnish institutions was considered important during the "long summer of migration" and the freezing of EU/Finnish-Russian relations in 2014. The national institutional stability provided a mechanism for coping with the geopolitical changes in the regional scale. Also, in comparison with the EU internal Finnish-Swedish border, the resistance to the COVID-19 border intervention has been much more modest in the Finnish-Russian borderland, where the cross-border traffic similarly decreased. It is possible that knowledge and experiences of sudden border restrictions and immobility at the securitized EU external borders have, in a way, prepared borderland communities to cope with and adapt to sudden border transitions, both decreasing resistance and increasing resilience with respect to the COVID-19 travel restrictions.

The narrative framing that connects the importance of increasing resilience with the notion of a world of permeable borders is ambiguous and tends to naturalize borders as rigid lines that divide communities and determine their future. Yet global challenges know no borders and cannot be overcome through bordering and border drawing (Dalby 2019). Environmental challenges like climate change need to be mitigated, coped with, prevented, and anticipated with the help of cross-border cooperation. In an ideal case, experienced crises and challenges would bound borderland actors together across the border to establish new connections and organizations of cooperation. This is a challenging task especially in geopolitically sensitive borderlands where tensions and conflicts of interests are present in many ways. The EU neighborhood policy provides a fitting example of how values and conflicts of interest influence and shape the conception of resilience. Also, European internal cross-border regions are often considered more "artificial" and thus more vulnerable to political turbulence than are the traditional state and sub-state regions (Perkmann 2002). A critical question is whether and to what extent cross-border regions and trust-based cross-border connections are resilient to border transitions and securitization. More knowledge is needed on how multilayered cross-border connections are vulnerable or resilient with regards to different political, economic, social and environmental changes and continuous border disruptions.

Note

1 The author conducted observation at the Finnish-Swedish border at the Victorian Square at the Tornio–Haparanda pedestrian border crossing point and at the Aavasaksa–Övertorneå border crossing point in March–June 2020.

References

Ackleson, J., 2005. Constructing security on the US-Mexico border. *Political Geography*, 24 (2), 165–184.

Allen, C., Angeler, D., Ahjond, S., Garmestani, L., Gunderson, L., and Holling, C.S., 2014. Panarchy: theory and application. *Ecosystems*, 17 (4), 578–589.

Amoore, L., 2006. Biometric borders: governing mobilities in the war terror. *Political Geography*, 25 (3), 336–351.

Andersen, D. and Sandberg, M. 2012. Introduction. *In*: D. Andersen, M. Klatt and M. Sandberg, eds. *The border multiple: the practicing of borders between public policy and everyday life in a RE-scaling Europe*. Farnham: Ashgate, 1–23.

Andersen, D.J., 2022. Line-practice as resilience strategy: the Istrian experience. *In*: D.J. Andersen and E-K. Prokkola, eds. *Borderlands resilience: transitions, adaptation, and resistance at borders*. London: Routledge, 166–181.

Anderson, J., 1996. The shifting stage of politics: new medieval and postmodern territorialities. *Environment and Planning: Society and Space*, 4, 133–153.

Anholt, R. and Sinatti, G. 2020. Under the guise of resilience: the EU approach to migration and forced displacement in Jordan and Lebanon. *Contemporary Security Policy*, 41 (2), 311–335.

Anzaldúa, G. 1987. *Borderlands/La Frontera: the new mestiza*. San Francisco, CA: Aunt Lute Books.

Baldersheim, H. and Ståhlberg, K., eds., 1999. *Nordic region-building in a European perspective*. Aldershot: Ashgate.

Bialasiewicz, L., 2012. Off-shoring and out-sourcing the borders of EUrope: Libya and EU border work in the Mediterranean. *Geopolitics*, 17 (4), 843–866.

Biscop, S., 2017. A strategy for Europe's neighbourhood: keep resilient and carry on? ARI 4/2017–16/1/2017. Available from: www.realinstitutoelcano.org/wps/portal/rielcano_en/contenido?WCM_GLOBAL_CONTEXT=/elcano/elcano_in/zonas_in/ari4-2017-biscop-strategy-europe-neighbourhood-keep-resilient-carry-on (accessed 30 December 2020).

Boschma, R., 2015. Evolutionary economic geography: towards an evolutionary perspective on regional resilience. *Regional Studies*, 49 (5), 733–751.

Bourbeau, P., 2015. Migration, resilience and security. *Journal of Ethnic and Migration Studies*, 41 (12), 1958–1977.

Bristow, G. and Healy, A., 2020. Supranational policy and economic shocks: the role of EU structural funds in the economic resilience of regions. *In*: G. Bristow and A. Healy, eds. *Handbook of regional economic resilience*. Cheltenham: Elgar, 280–298.

Christopherson, S., Michie, J., and Tyler, P., 2010. Regional resilience: theoretical and empirical perspectives. *Cambridge Journal of Regions, Economy and Society*, 3 (1), 3–10.

Cote, M. and Nightingale, A., 2012. Resilience thinking meets social theory: situating social change in socio-ecological systems (SES) research. *Progress in Human Geography*, 36 (4), 475–489.

Cutter, S.L., 2016. Resilience to what? Resilience for whom? *Geographical Journal*, 182, 110–113.

Dalby, S., 2019. Unsustainable borders?: climate geopolitics in a warming world. Paper for presentation to a workshop on 'New Directions at the Border'. Ottawa: Carleton University, 17 October 2019.

Eickhoff, K. and Stollenwerk, E., 2018. Strengthening resilience in the EU's neighbourhood. EU-LISTCO Policy Paper Series No. 01. Available from: https://static1.squarespace.com/static/5afd4286f407b4a0bd8d974f/t/5cffaacdd8b3cc000151aa1d/1560259278401/EU-LISTCO+POLICY+PAPERS_01+Eickhoff%26Stollenwerk.pdf (accessed 30 December 2020).

EU, 2016. Shared vision, common action: a stronger Europe. A global strategy for the European union's foreign and security policy. Available from: https://eeas.europa.eu/archives/docs/top_stories/pdf/eugs_review_web.pdf (accessed 17.9.2021).

European Commission and High Presentative of the Union for Foreign Affairs and Security Policy, 2015. Review of the European Neighbourhood Policy. Brussels, 18.11.2015 JOIN(2015) 50 final. Available from: https://ec.europa.eu/neighbourhood-enlargement/sites/near/files/joint-communication_review-of-the-enp.pdf (accessed 30 December 2020).

Finnish Border Guard, 2020. News and bulletin [Uutiset ja tiedotteet]. Available from: https://raja.fi/uutiset-ja-tiedotteet/-/asset_publisher/kBNrdPA9Hj7T/content/id/52904008# (accessed 30 December 2020).

Frandsen, S.B., 2022. Schleswig. From a land-in-between to a national borderland. *In*: D.J. Andersen and E-K. Prokkola, eds. *Borderlands resilience: transitions, adaptation, and resistance at borders*. London: Routledge, 121–136.

Hannonen, O., 2022. Mobility turbulences and second-home resilience across the Finnish-Russian border. *In*: D.J. Andersen and E-K. Prokkola, eds. *Borderlands resilience: transitions, adaptation, and resistance at borders*. London: Routledge, 90–105.

Hassink, R., 2010. Regional resilience: a promising concept to explain differences in regional economic adaptability? *Cambridge Journal of Regions, Economy and Society*, 3 (1), 45–58.

Jensen, O. and Richardson, T., 2004. *Making European space: mobility, power and territorial identity*. London: Routledge.

Juntti, P., 2020. 'Tässä liikutaan harmaalla alueella' – Hallitus kielsi suomalaisten rajanylitykset koronaviruksen takia, mutta asiantuntijan mukaan kielto ei perustu lakiin. *Helsingin Sanomat*. Available from: https://www.hs.fi/kotimaa/art-2000006497325.html (accessed 06 May 2020).

Koch, K., 2018. Geopolitics of cross-border cooperation at the EU's external borders. *Nordia Geographical Publications*, 47 (1).

Lois, M., Cairo, H., and de las Heras, G., 2022. Politics of resilience . . . politics of borders? In-mobility, insecurity and Schengen "exceptional circumstances" in the time of COVID-19 at the Spanish-Portuguese border. *In*: D.J. Andersen and E-K. Prokkola, eds. *Borderlands resilience: transitions, adaptation, and resistance at borders*. London: Routledge, 54-69.

Martinez, O., 1994. The dynamics of border interaction: new approaches to border analysis. *In*: C. Scholfield, ed., *Global boundaries*. London: Routlegde, 1–15.

Meerow, S. and Newell, J., 2019. Urban resilience for whom, what, when, where, and why? *Urban Geography*, 40 (3), 309–329.

Newman, D. and Paasi, A., 1998. Fences and neighbours in the postmodern world: boundary narratives in political geography. *Progress in Human Geography*, 22 (2), 186–207.

O'Dowd, L., 2010. From a "borderless world" to a "world of borders": 'bringing history back in. *Environment and Planning D: Society and Space*, 28 (6), 1031–1050.

Opiłowska, E., 2021. The COVID-19 crisis: the end of a borderless Europe? *European Societies*, 23 (1), 589–600.

Paasi, A., 2003. Region and place: regional identity in question. *Progress in Human Geography*, 27, 475–485.

Paasi, A. and Prokkola, E-K., 2008. Territorial dynamics, cross-border work and everyday life in the Finnish – Swedish border area. *Space and Polity*, 12 (1), 13–29.

Paasi, A., Prokkola, E-K., Saarinen, J., and Zimmerbauer, K., eds., 2019. *Borderless worlds for whom? Ethics, moralities and mobilities*. London: Routledge.

Passoja, A., 2020. Teräsaitoja, huonoa kohtelua ja perhetapaamisia aidan yli: näin sisärajoilla on eletty poikkeusaikaa, joka on venynyt jo laittoman pitkäksi. *Yle Uutiset*. https://yle.fi/uutiset/3-11657563 (accessed 23 November 2021).

Perkmann, M., 2002. Euroregions: institutional entrepreneurship in the European Union. *In*: M. Perkmann and N. Sum, eds. *Globalization, regionalization and cross-border regions*. Hampshire: Palgrave Macmillan, 103–124.

Phelan, L., Henderson-Sellers, A., and Taplin, R. 2013. The political economy of addressing the climate crisis in the earth system: undermining perverse resilience. *New Political Economy*, 18 (2), 198–226.

Prokkola, E-K., 2019. Border-regional resilience in EU internal and external border areas in Finland. *European Planning Studies*, 27 (8), 1587–1606.

Prokkola, E-K., 2020. Geopolitics of border securitization: sovereignty, nationalism and solidarity in asylum reception in Finland. *Geopolitics*, 25 (4), 867–886.

Prokkola, E-K. and Lois, M., 2016. Scalar politics of border heritage. *Scandinavian Journal of Hospitality and Tourism*, 16 (1), 14–35.

Rogers, P., Bohland, J., and Lawrence, J., 2020. Resilience and values: global perspectives on the values and worldviews underpinning the resilience concept. *Political Geography*, 83.

Sahlins, P., 1989. *Boundaries: the making of France and Spain in the Pyrenees*. Oxford: University California Press.

Scheel, S., 2015. Das Konzept der Autonomie der Migration überdenken? – Yes Please! [Rethinking the concept of autonomy of migration? – Yes Please!]. *Movements*, 1 (2), 1–15.

Simon, S. and Randalls, S., 2016. Geography, ontological politics and the resilient future. *Dialogues in Human Geography*, 6 (1), 3–18.

Sohn, C., 2014. Modelling cross-border integration: the role of borders as resource. *Geopolitics*, 19 (3), 587–608.

Statistic Finland, 2020. Accommodation statistics from 2013 and 2016. Available from: www.stat.fi/til/matk/index_en.html (accessed 30 December 2020).

Svensson, S., 2015. The bordered world of cross-border cooperation. *Regional & Federal Studies*, 25 (3), 277–295.

Timothy, D., 1999. Cross-border partnership in tourism resource management: international parks along the US-Canada border. *Journal of Sustainable Tourism*, 7 (3–4), 182–205.

van Houtum, H., 2000. III European perspectives on borderlands. *Journal of Borderlands Studies*, 15 (1), 56–83.

van Houtum, H., Kramsch, O., and Zierhofer, W., 2005. Prologue b/ordering space. *In*: H. van Houtum, O. Kramsch and F. Zierhofer, eds. *B/ordering space*. Aldershot: Ashgate, 1–13.

Wagner, W. and Anholt, R., 2016. Resilience as the EU Global Strategy's new leitmotif: pragmatic, problematic or promising? *Contemporary Security Policy*, 37 (3), 414–430.

Walker, B., Holling, C.S., Carpenter, S.R., and Kinzig, A., 2004. Resilience, adaptability and transformability in social – ecological systems. *Ecology and Society*, 9 (2).

Wandji, G., 2019. Rethinking the time and space of resilience beyond the West: an example of the post-colonial border. *Resilience, International Policies, Practices and Discourses*, 7 (3), 288–303.

Welsh, M., 2014. Resilience and responsibility: governing uncertainty in a complex world. *The Geographical Journal*, 180 (1), 15–26.

3 Cross-border resilience in higher education

Brexit and its impact on Irish–Northern Irish university cross-border cooperation

Katharina Koch

Introduction

On June 23, 2016, a referendum was held on the withdrawal of the United Kingdom (UK) from the European Union (EU). The majority of those who voted to leave the EU was narrow (51.9%) but resulted in a tremendous effort of negotiating the exit of the UK when Article 50 of the Lisbon Treaty was triggered on March 29, 2017. The formal departure of the UK from the EU was completed on January 31, 2020, followed by a transition period to negotiate a trade agreement, which was concluded on December 31, 2020 (European Commission 2020a). The prolonged uncertainty of the negotiations generated tremendous uncertainty across various policy domains that are of relevance to both the UK and the EU.

Researchers have investigated the impact of Brexit on various policy sectors that are dependent on the mobility of goods, capital and people, reflecting the three pillars of the European Single Market which Brexiteers had aimed to repeal. The effects of Brexit have been studied in a variety of contexts such as the Irish peace process (Hayward and Murphy 2018), cross-border collaboration (O'Keeffe and Creamer 2019) and cross-border civil service cooperation (Tannam 2018). Nevertheless, scholars have criticized that the hitherto-frictionless Irish–Northern Irish border had been mentioned rarely during pre-referendum debates involving the public, thus undermining its status as a stabilizing factor in the region (Burke 2016).

This was however not the case for public officials and policymakers who work in sectors that depend on cross-border mobility. An early impact of Brexit was that higher education (HE) institutions started to prepare contingency plans for a variety of Brexit scenarios, including the possibility of a hard border between Northern Ireland and the Republic of Ireland.[1] This chapter focuses on cross-border cooperation (CBC) between universities and the implications of Brexit for students and academic staff in Ireland and Northern Ireland (UK). The analytical focus is on the concept of *cross-border mobility*, which Prokkola (2019) argues is an important resource that enhances regional resilience. Similarly, Korhonen et al. (2021) argue that CBC is crucial for establishing regional resilience.

DOI: 10.4324/9781003131328-4

Inspired by Rumford's (2012) "multiperspectival" conceptualization of borders, the Irish–Northern Irish border is understood from a relational perspective (also see Paasi 2012). Rumford (2012, p. 899) argues that "borders can be located 'away from the border' and dispersed throughout society." In the context of HE CBC, cross-border mobility does therefore not only refer to the physical movement of students and academics across the border but also considers the impact of Brexit on research collaborations and student exchanges. For example, although the trade agreement between the EU and the UK allows British universities to apply for the European Research Council's (ERC) funding stream, the EU-funded Erasmus student exchange scholarship is no longer an option for UK students. Thus, the case study of HE will operationalize resilience as the adapted and renewed strategies developed by universities in both Ireland and Northern Ireland to maintain mobility. These strategies aim to address the uncertainty of the Brexit process by preparing several contingency plans for various scenarios in the HE sector and the potential disruption of student, faculty and staff mobility. Resilience is conceptualized from the perspective of mobility, especially the response to threats that could impact frictionless cross-border mobility.

The uncertainty regarding the future relations of HE between Northern Ireland and Ireland is the main focus in this chapter, which will answer the following questions: (1) How does HE mobility contribute to cross-border regional resilience between Ireland and Northern Ireland? (2) How is the border represented in the HE mitigation strategies and plans that address the uncertainty of Brexit? (3) How do HE actors address possible future frictions and tensions regarding the border? The first section discusses the Irish–Northern Irish border case by examining how Brexit impacts cross-border stability. The second section offers a conceptualization of resilience in the context of HE cross-border mobility and also offers a brief methodological overview. The third section discusses the implications of Brexit for HE cross-border mobility.

The Irish–Northern Irish border case

It is the unique spatial and geographical position of the island of Ireland and the historical and social particularities that have shaped the relationship between Northern Ireland and Ireland until nowadays. As discussed by Andersen and Prokkola (2022) in the Introduction, resilience needs to be interpreted contextually, and in this chapter it means taking the cultural, political and geo-historical contexts of the Irish–Northern Irish border into consideration in order to make sense of HE mobility as a form of borderland resilience. Undoubtedly, the Brexit negotiations were greatly influenced by the Irish–Northern Irish border debate. The UK wanted to ensure that Northern Ireland remains a full member of the UK by maintaining free travel, market access and a tariff-free relationship with the country. Yet, the Irish channel divides Northern Ireland from Great Britain, rendering it a direct neighbor with Ireland in the south, which is a member of the EU. The EU aimed to maintain the seamless and frictionless connection between both countries by possibly drawing a line in the Irish sea – something that the

UK wanted to avoid at all costs. In this context, Hayward (2018, p. 239) argues that "any significant shift in the status of the border has consequences not just for Northern Ireland but for the peace process more broadly."

From the 1960s until the late 1990s, Northern Ireland has been a site of an ethno-nationalist conflict, sometimes described as a "guerrilla war" that was defined as a political and nationalistic conflict with ethnic and sectarian dimensions, a conflict claiming the lives of more than 3,500 people, half of them civilians. In 1998, the Good Friday Agreement was signed to finally resolve the conflict, and the agreement together with a wide range of peace amelioration initiatives has maintained peace during the last two decades on the island. The border between Ireland and Northern Ireland has been opened for people's traffic since 1953, but militarized border checkpoints were established during the 1970s and removed after the Good Friday Agreement. Ireland and the UK became members in the European Single Market in 1993, de facto removing any remaining barriers for cross-border travel of both people and goods with the rest of the EU.

The peace process and the open-border arrangements were threatened by the prospect of a hard border due to Brexit (Hayward and Murphy 2018). The Brexit referendum jeopardized not only the continuous peace process since the Good Friday Agreement but also created uncertainty among the population regarding the future status of the border. Especially in the context of HE, students, staff and faculty on both sides of the border were facing uncertainty regarding funding and tuition fee arrangements. While Kolossov (2005, p. 619) argues that borders configured as barriers are "not only inefficient but objectively harmful to society and the economy," other scholars emphasize that borders "are no longer understood merely as barriers, but also as resources, bridges, and points of contact" (Prokkola 2011, p. 1190). Cross-border mobility has become an important resource for both Ireland and Northern Ireland since the Good Friday Agreement, not to mention how the frictionless cross-border mobility and CBC in the HE context significantly contribute to the continuous Irish peace process.

Cross-border regional resilience and HE mobility

As stated by Andersen and Prokkola (2022), borders and borderlands are transforming. In the Irish context, Brexit has disrupted the HE sector as the hitherto-frictionless academic mobility between the UK and Ireland was threatened after the referendum. In order to make sense of the Irish and Northern Irish responses to Brexit in the HE sector, this chapter discusses resilience based on the relevance of cross-border mobility for HE institutions on the island of Ireland.

According to Salter (2013), mobility has been studied in a variety of contexts, including territoriality (Rumford 2006), border studies (Prokkola 2019) as well as resilience in cross-border regions (Slusarciuc 2017). Resilience as a concept remains ambiguous as it has been adopted by a variety of academic disciplines. In human geography, resilience has "come to stand for the ability to absorb, withstand, persist, thrive and reorganize in the face of the shocks and disturbances of always uncertain becoming" (Simon and Randalls 2016, p. 4). Yet, this general definition remains rather vague as it does not hint towards specific factors that contribute to resilience. Thus, as Andersen and Prokkola (2022) point out, the application of resilience as a theory is a challenge as it needs to be meaningfully adapted to the specific field of study, context and research questions. For example, mitigation strategies and contingency plans vary across policy domains and are tailored for specific purposes, such as HE, and they can significantly differ from other policy sectors, such as foreign policy and security.

Discussions across the humanities and social sciences usually apply the concept of resilience in the context of "crisis" (Malkki and Sinkkonen 2016), "shocks" or "shock absorption" (see Alessi et al. 2019), or "adaptation" as a coping mechanism (Marque and Miralles-Guasch 2018). In the case of cross-border regions, it would be simplistic to deduce that a well-integrated border is immune to shocks and disturbances. Even a seemingly well-integrated and prospering border region such as the one between Germany and France has inherited "deep vulnerabilities" maintaining cultural distances, occasional mistrust and other barriers impeding borderlands resilience (Adrot et al. 2018, p. 430).

In general, HE institutions are considered gatekeepers of formalized knowledge and represent institutions crucial for the maintenance of economic competitiveness. In fact, HE institutions can be considered key actors in relation to the four freedoms of mobility in the EU, particularly supporting the Lisbon Treaty's aim of generating a knowledge-based integrated economy and society (Boyer 2009). HE institutions are usually associated with an increase of personal and community standards of living by raising the level of education among populations. Thus, universities and other HE institutions have become important elements in the global development strategies that result in an ever-expanding volume of cross-border activities. Furthermore, and this is unique to the cross-border relationship between Ireland and Northern Ireland, Komarova (2017, p. 2) shows that HE has become a significant contributor to the Irish peace process and an all-island economy. The Department for the Economy (2012, p. 23) in Northern Ireland explicitly states that:

> Cross-border co-operation and undergraduate mobility between institutions in Northern Ireland and the Republic of Ireland are important from an economic, social and cultural perspective. Such activity has the potential to drive the growth of the all-island economy, broaden the pool of graduates both north and south and support the ongoing peace process. The Department is committed to supporting cross-border co-operation in teaching and learning, with the aim of increasing understanding, sharing good practice and enabling students to move freely between the two jurisdictions.

Similarly, Ireland also supports cross-border student mobility, particularly between North and South (Higher Education Authority 2011). In this way, it is feasible to argue that HE contributes to the peace-building process by fostering cross-border relations across a border which has experienced significant conflict during the 20th century. Another compelling example in which HE can contribute to cross-border regional stability is the Erasmus exchange program funded by the EU. Mitchell (2012) argues that the Erasmus program is a "civic experience" promoting cultural awareness and the recognition of different identities across Europe while strengthening the idea of Europe and support for the EU as a supranational institution.

The potential disruption of HE cross-border mobility between Ireland and the UK through Brexit can be understood as a "crisis" which threatens the process of conflict amelioration and peace-building that is significantly fostered through CBC and cross-border mobility (see McCall 2013). Open borders and CBC are considered an important resource for the economic and social wellbeing of the population. Prokkola (2019, p. 1601) argues that open border regimes "increase regional resilience in many ways, especially in terms of cross-border networking, information sharing and trust." Prokkola (2019) offers an analysis of mobility/immobility shocks at the Finnish borders with Sweden and Russia. She shows that crisis management moves through various stages in which cross-border actors exchange information and mobilize formal and informal networks to adapt to the changing situation by, for example, establishing new models of rapid-response systems.

Brexit as a crisis must be contextualized as one element in the "polycrisis" that has affected the all-island economy. Ireland was already greatly impacted during the 2007/2008 financial crisis as it lost its status as the "Celtic Tiger" in the European Single Market (Murphy and O'Brennan 2019) and entered a phase of economic austerity and reduced public expenditure (Pritchard and Slowey 2017). The Irish HE sector is still responding to these past disruptions, and the Brexit response is strategized in the context of prevailing stringent HE budgets. Brexit triggered a crisis response to the extent that HE institutions in both Ireland and Northern Ireland struggled to find adequate answers for students and staff worried about their future, either regarding tuition fee increases or residency status (CCT Dublin 2016). Furthermore, Brexit was not a rapidly evolving crisis; rather, negotiations were slow and tied to decision-making processes in both London and Brussels – thus, the crisis was relational in the sense that it involved several actors and competing interests, mostly between the negotiation parties of the UK and the EU.

The following analysis is based on policy documents and interviews with university representatives in Ireland.[2] The documents regarding the Brexit negotiations consist of a range of EU Commission documents[3] and 16 reports published by the Brexit Task Force of the Royal Irish Academy (RIA 2020). The Royal Irish Academy (RIA), formed in 1785, is an all-island academic institution, representing universities from both Ireland and Northern Ireland and across the sciences, humanities and social sciences. It has the goal to foster research excellency and to build a bridge between academia and society. The second part of the research material consists of ten semi-structured interviews held in the summer of 2018 with representatives of the European Commission Representation in Ireland

(Dublin) and with key actors at Trinity College Dublin (TCD), including an RIA representative. From a methodological perspective, it must be recognized that the interviews were carried out in Ireland; thus, concrete contingency plans are applicable in the context of TCD Dublin.

The material was analyzed using a theoretically informed content analysis (Hay 2016) to screen the material with respect to the conceptual framework that operationalizes cross-border mobility as a form of resilience. The goal was to identify the adopted strategies by HE institutions to organize future relations between Ireland and Northern Ireland. The speech built around the border and mobility in the context of HE forms the key analytical lens under which the concept of resilience is investigated. The analysis assumes that a "dialectical relationship" exists between political discourses, speech acts and statements and the institutions and social structures in which they are embedded (Koch 2018a, p. 57). This relationship is particularly relevant with regard to the study of Brexit because the strategies and scenarios imagined by the HE institutions also represent a reality that was produced unintentionally due to the uncertainty of the outcome of the negotiations and resulting lack of information prior to the Trade and Cooperation Agreement between the EU and the UK.

Brexit and HE CBC between Ireland and Northern Ireland

In response to Brexit, the RIA established a "Brexit Taskforce" in 2015 in order to develop strategies that address the possible implications of Brexit for the HE sector in both Northern Ireland and Ireland (RIA 2020). The task force was already established before the referendum in June 2016, and initial publications offer advice on potential future HE funding scenarios in Ireland in the context of both, the continuous recovery after the 2008 financial crisis (RIA 2016a) and the potential implications of the UK's referendum on EU membership (RIA 2016b). RIA conducted surveys with the academic community in both Ireland and Northern Ireland during 2016 and 2017 which established that, in general, Brexit may actually offer certain opportunities for Irish HE institutions; however, there was a common agreement among survey respondents that there were no benefits to be gained from Brexit for Northern Irish HE institutions. The following table shows the potential challenges for HE institutions in Northern Ireland while listing perceived opportunities and challenges for Ireland.

Despite these differing views, there was the common assumption among survey respondents that a closed border is undesirable for both countries. Respondents called for "academic and research mobility to be protected and retention of the Common Travel Area" (RIA 2017a, p. 8). Furthermore, a common opinion among respondents was that the preservation of existing collaborative research and academic arrangements would be the most important task for HE institutions going into post-Brexit scenario planning.

The immediate crisis response among universities, such as TCD or Queen's University in Belfast, was to raise awareness for the potential dangers of Brexit

Table 3.1 Impact of Brexit on Northern Ireland and opportunities and challenges for Ireland based on RIA (2016b) and RIA (2017a). Note that survey results are based on the information respondents had in 2016/17.

Potential impact of Brexit for HE in Northern Ireland	Potential opportunities and challenges for HE in Ireland
Limited access to EU funding programs.	Potential to gain more EU funding if UK universities become ineligible.
Threatens researcher's ability to participate in large-scale European research infrastructure.	Could disrupt Irish–Northern Irish collaboration enabled through EU funding.
Loss of talent as students and academics are drawn to Irish universities.	Attracting more international students and academics.
Uncertainty to which extent binational research cooperation with the UK could make up for the loss of EU funding.	Need for new funding schemes between Ireland and Northern Ireland/Great Britain.
Negative impact on local economy that is dependent on the public sector.	Strengthening of local economy if more international students arrive (housing etc.) but could also strain capacities in, for example, Dublin (risk of overcrowding).
EU students could pay higher tuition fees; however, may discourage international students (including Irish students) + challenges to recruit overseas students due to potential hard borders.	Irish students may be discouraged to study in Northern Ireland and Great Britain due to tuition fees increases and thus study domestically (loss of exchange experience).
Loss of exchange students (i.e. Erasmus program).	Potentially higher success rates in EU's Marie Curie Sklodowska Actions and Eramus+ programs.

for students and academics and the collaboration between Ireland and the UK. The Provost of TCD warned that the vote "will have a long-term impact on universities in the Republic of Ireland and could hinder the long tradition of students moving between the islands as well as important research activity" (CCT Dublin 2016). In this way, the establishment of the RIA Task Force showcases an important step in Ireland's and Northern Ireland's crisis management approach. At the same time, the task force and other HE actors immediately raised awareness for a potential disruption of Irish–Northern Irish relations, emphasizing the vulnerability of the region to shocks that could disrupt regional stability. For example, Dublin City University (DCU) formed the Brexit Institute in 2016, aiming to offer a "leading platform to document and debate developments in the relations between the UK and the EU" (DCU 2017). The goal was to actively engage with the public and to raise awareness for the importance of HE CBC between Ireland and Northern Ireland (as well as with the broader UK). Queen's University in Belfast established "Queen's on Brexit" to inform the Northern Irish public through a body of experts while also offering advice for prospective and current students (Queen's University Belfast 2021).

What followed the initial shock response in 2016 was a phase of careful research in the form of surveys and other research methods. It can thus be argued that research and knowledge, as well as anticipation and informed predictions, are important elements of building resilience in HE CBC. As an immediate response to Brexit, enhanced communication took place between Irish and Northern Irish HE officials and policymakers for the purpose of communicating survey results and to offer policy advice, especially for the EU Commission in Brussels. Therefore, the crisis management approach moved into the stage of concrete planning. RIA and Irish/Northern Irish HE institutions developed several contingency plans and long-term adaptation strategies, based on various Brexit scenarios (open/closed/hybrid border models) that focused on student and staff retainment, new funding opportunities, collaboration with existing European research networks and preparation for new administrative hurdles regarding cross-border travel. These contingency plans are a robust sign of existing cross-border resilience as existing connections, underlined by institutional and personal trust relations, foster CBC. The common goal was to offer clarity to students and staff while limiting potentially damaging consequences of Brexit for cross-border HE collaboration, including the maintenance of funding opportunities.

While RIA was preparing for a variety of open or closed border scenarios, all adaptation strategies had however the tendency to alter the status quo in an irreversible way (Alonso et al. 2019). For example, since the Brexit referendum, Irish universities have re-thought future collaboration with the European continent by focusing on establishing stronger ties with universities outside the UK as it was stated by a TCD official during an interview. The focus of Irish universities thus started to shift towards entering academic research networks on the European continent; yet, academic collaboration with the UK remains strong (Figure 3.1). Furthermore, the research collaboration between TCD and the Queen's University in Belfast is significant as it represents the fourth position behind University College London, CNRS in France and King's College London in terms of joint academic publications.

The strong research collaboration between Ireland and Northern Ireland (as well as with Great Britain) was an important factor influencing Brexit negotiations, thus showcasing that cross-border academic cooperation significantly contributes to regional resilience. The close ties between Irish and Northern Irish HE institutions foster cross-border research collaboration and thus also support efforts of developing contingency measures to off-set any negative impact of Brexit on research and education.

The literature on resilience does not offer one single definition that would capture the multiple ways resilience is used to navigate contemporary crisis politics. Resilience in HE is an important feature but at the same time reflects a process in the sense that it develops over time, especially through trust relations both on personal levels but also between institutions (see Koch 2018b). Cross-border regional resilience stems from a variety of strengths, for example effective CBC,

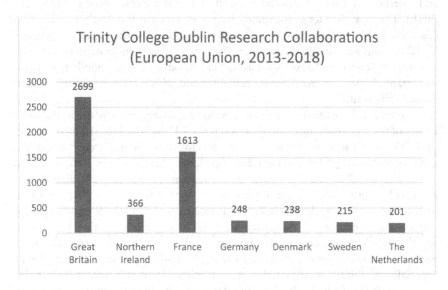

Figure 3.1 Co-authored research papers 2013–2018.

Note: The columns are divided in the following: Great Britain (nine universities); Northern Ireland (one university); France (six universities); Germany, Denmark, Sweden, the Netherlands (one university respectively).

Source: TCD 2020.

cross-border mobility, trust in institutions (Jakola and Prokkola 2017) as well as the recognition among the public that an open border and free travel are crucial for the prosperity of the border region (Prokkola 2019).

This can translate into the development of shared values within a society. In the HE context, the EU's Erasmus exchange program is an excellent example of instilling the shared value of free cross-border mobility among young and highly educated citizens. One crucial issue, as detected by a TCD representative, is that "we share the same culture and language in the Republic of Ireland as with the UK and thus, naturally, our goal is to strengthen our ties with the UK." Therefore, shared values are directly interlinked with resilience (Rogers et al. 2020). Yet, further strengthening of these important ties between the two countries could be greatly affected by Brexit in the long run.

Some scholars argue that the UK is already experiencing a "brain drain" as EU citizens, including Irish, are moving or refusing to apply for British universities as they fear to lose the right to live in Great Britain or Northern Ireland (Cressey 2016). Therefore, even though both Great Britain and Northern Ireland still remained for four years in the European Single Market since 2016, the "mental borders" among the Irish, the Northern Irish and the British in the HE sector were erected almost immediately when the results of the Brexit vote were announced. This indicates a re-bordering phenomenon not only between Ireland

and Northern Ireland but also between the UK and the rest of the EU (McCall 2018). The new "border in the mind" (Gormley-Heenan and Aughey 2017) can cause the public to lose trust in public institutions across the border, especially if HE institutions are not clearly conveying their mitigation strategies to off-set any negative effects of Brexit, such as higher tuition fees. Thus, public trust is an important element of cross-border resilience and which HE institutions should foster in the future.

Currently, HE institutions are still developing strategies to adapt to the new reality of the free trade agreement between the EU and the UK. An interviewee in TCD emphasized in 2018 that any developed strategies were "based on possible scenarios and not on certain knowledge. Yet, it affects staff, students and the college." Brexit is considered to have a major impact on collaborative projects because from a research perspective, the most important aspect is the ability to move, re-settle or relocate. Although the COVID-19 pandemic has proven that remote work, especially in the HE sector but also other industries, is efficient, there are certain circumstances where in-person classes and net-working are crucial for the progress of research, science and knowledge. This is especially valid in the medical fields, natural sciences and life sciences but also in the social sciences which often require fieldwork and interactions with research participants. Therefore, a physical connection remains crucial, and this aspect is used as a key feature in RIA's post-Brexit scenario planning. RIA (2017b, p. 4) developed three post-Brexit strategic actions for HE in Northern Ireland and Ireland:

1 Actions to maintain the beneficial north-south, east-west axis between Ire-land and Northern Ireland, and Ireland and Britain, in higher education and research.
2 Actions to address underinvestment in HE (North and South) and research to better position Ireland as a global hub for excellence in research, teaching and learning.
3 Actions to align national research and internationalization strategies that will grow Ireland's international research connectivity and enhance its reputation as a hub for international talent.

Besides institutional aspects, resilience also develops through personal creativ-ity and ability to establish new ways of maintaining cross-border collaboration and to adapt to a "new normal." In this way, not only HE institutions but also academics, researchers and students are developing ways, within the constraints of their personal, financial and social circumstances, to adapt to a situation in which cross-border mobility could be affected after the exit of the UK from the European Single Market. However, major support and back-up must come from the HE institutions directly. For example, TCD became a member of the League of European Research Universities (LERU) in January 2017. LERU is a consor-tium of European universities with the purpose of lobbying for close HE collabo-ration on the EU level. On January 31, 2020, the day that the UK formally left the EU, LERU published a statement advocating for the continued collaboration between EU and UK universities (LERU 2020).

Another often-quoted issue is related to HE infrastructure and the current inability of Irish universities to accommodate for the expected large influx of EU and US students once Great Britain and Northern Ireland are no longer integrated into the funding scheme of the European Commission and student scholarships, such as Erasmus and Erasmus+. Ireland will effectively represent the only remaining anglophone country in the EU; thus, it is expected that it will draw large numbers of foreign students and staff due to the anglophone dominance in academia. While British universities will remain attractive as student destinations, EU citizens especially will re-direct their interest towards Ireland for the ease of access and freedom of mobility within the EU. This once more underlines the importance of frictionless cross-border mobility in HE. The future border scenario between Ireland and Northern Ireland will determine to which extent HE crisis planning, and lobbying, were successful and contributed to the stabilization of cross-border relations.

Conclusion

The concept of resilience, studied in the context of HE cross-border mobility on the island of Ireland and between Ireland and the UK, is a useful analytical tool to understand the crisis management approach of universities and other HE institutions, such as RIA. This chapter shows that cross-border mobility is an important factor contributing to the regional stability between Ireland and Northern Ireland and, furthermore, that Brexit could greatly impact the sensitive peace that was accomplished with the Good Friday Agreement in 1998. HE mobility can be acknowledged as an important aspect in the peace process as student exchanges and academic cross-border collaboration raise cultural awareness and understanding for different national identities. This is particularly important in the Irish context, which has experienced decades of violent struggle during the 20th century.

From a theoretical perspective, this case study shows the importance of recognizing borders and borderlands from a multiperspectival approach (Rumford 2012) which underlines that bordering processes are produced through multiple bordering practices that can happen "away from borderlines" (Komarova and Hayward 2019, p. 545). This means that border/lands resilience is not limited to practices at the physical border but are also reflected in the efforts, strategies, policies, negotiations and speeches to maintain for example frictionless cross-border mobility. Borderlands resilience should thus also be understood from a relational perspective (see also Prokkola 2021). This means that power relations and negotiations, for example between actors with different interests, influence the resilience of a border to shocks and disturbances. In the case of Brexit and its impact on the Irish–Northern Irish borderlands, the progress of negotiations between London and Brussels was crucial for HE practitioners to develop and adjust their contingency plans and preparatory strategies.

Since the results of the Brexit referendum were announced, HE institutions in both Ireland and Northern Ireland have immediately begun to develop contingency plans and to prepare mitigation strategies for various post-Brexit scenarios. The Irish border was a dominant aspect in all plans as academics emphasized the importance of continued Irish–Northern Irish and Irish-UK collaboration. As shown in the analysis, British universities and Northern Ireland's Queen's University in Belfast are among the most important institutions with whom Irish researchers at TCD collaborate. Furthermore, it was emphasized that frictionless travel must be preserved, especially on the island of Ireland, not only for the purpose of academic cooperation but also to foster peaceful relations. In this way, HE takes on a geopolitical role as a regional stabilizer.

The resilience shown by academics and HE institutions on the island of Ireland stems from and is illustrated by their immediate crisis response capabilities and the strong support from other EU universities and researchers that were mobilized to lobby in Brussels for a continued open border regime on the island of Ireland. This shows the relational aspect of resilience, meaning that borderland resilience does not only stem from a bounded region or territory but can also be fostered through cross-border institutional interactions. However, at the same time, the crisis response that was triggered by Brexit in Ireland and Northern Ireland also underlines the vulnerability of the HE sector and its dependence on free cross-border movement, particularly in the case of Northern Ireland for which academics could not find any advantages of a possible no-deal Brexit scenario. From an Irish perspective, some opportunities were mentioned related to higher student numbers as well as potential higher success rates in grant funding competitions administered by the EU.

The analysis also shows that HE institutions are not the only actors developing resilience to adverse scenarios or shocks/disruptions. Interviewers also highlighted that individual academics and researchers are currently developing their own ways to maintain research collaborations and international student recruitment. However, the common tenet is that academics would have wished to maintain the status quo prior to Brexit. Furthermore, Brexit can be argued to have turned into a self-fulfilling prophecy. Although in 2016 the UK was still a full member in the EU and the Single Market, HE institutions began to prepare for the worst-possible outcome, which would be a hard border between Ireland and Northern Ireland as well as a no-deal Brexit. Since these strategies are partially still underway, i.e. the identification of new research networks and funding streams on the European continent, Brexit had already started to transform the Irish HE sector since the day the referendum results were announced.

Nevertheless, RIA's contingency strategies always emphasize the importance of continued free movement across the Irish border, and thus the border itself has transformed into a site and form of resilience that could continue stabilizing the hitherto-sensitive relationship between Ireland and Northern Ireland – especially if the open border policy between Ireland and Northern Ireland remains intact. In the same way, a hardening of the border, for example with regard to future trade regulations, could undermine the two decades of progress that the Irish peace-building has accomplished since 1998 and the Irish border could – once more – erupt into a region of conflict. Continued HE frictionless cross-border mobility could prevent such a scenario while strengthening Ireland's border region resilience.

Notes

1 For the remainder of this chapter, "Ireland" refers to the "Republic of Ireland" to emphasize its distinction from the country of "Northern Ireland," which is part of the United Kingdom.
2 The research for this chapter lasted from 2018 until the end of 2020. Although the chapter makes every effort to reflect on current relations between the EU and UK, the author recognizes that the Brexit negotiations were volatile and will continue to define future EU-UK relations also in the following years. The interviews were held during 2018, but the analyzed policy documents include the period between the referendum results in 2016 until the EU-UK Trade and Cooperation Agreement finalized on December 31, 2021 (European Commission 2020a).
3 These documents are divided into the following: 144 until January 1, 2020 (European Commission 2020b) and 66 documents after January 1, 2020 (European Commission 2020c).

References

Adrot, A., Friedrich, F., Lotter, A., Münzberg, R.E., Wiens, M., and Raskob, W., 2018. Challenges in establishing cross-border resilience. *In*: A. Fekete and F. Fiedrich, eds. *Urban disaster resilience and security*. Cham: Springer international Publishing, 429–460.

Alessi, L., Benczur, P., Camolongo, F., Cariboni, J., Manca, A.R., Menhyert, B., and Pagano, A., 2019. The resilience of EU member states to the financial and economic crisis. *Social Indicators Research*, 18, 569–598. Available from: https://link.springer.com/article/10.1007/s11205-019-02200-1 (accessed 18 March 2021).

Alonso, A.D., Kok, S., and O'Brien, S., 2019. Uncertainty and adaptation in the context of Brexit: an entrepreneurial action and dynamic capabilities approach. *European Business Review*, 31 (6), 885–909. Available from: www.emerald.com/insight/content/doi/10.1108/EBR-05-2018-0101/full/html (accessed 18 March 2021).

Andersen, D.J. and Prokkola, E-K., 2022. Introduction: embedding borderlands resilience. *In*: D.J. Andersen and E-K. Prokkola, eds. *Borderlands resilience – transitions, adaption and resistance at borders*. London: Routledge, 1–18.

Boyer, R., 2009. From the Lisbon agenda to the Lisbon treaty: national research systems in the context of European integration and globalization. *In*: H. Delanghe, U. Muldur and L. Soete, eds. *European science and technology policy: towards integration or fragmentation*. Gloucestershire: Edward Elgar Publishing Limited, 101–126.

Burke, E., 2016. Who will speak for Northern Ireland? *The RUSI Journal*, 161 (2), 4–12. Available from: www.tandfonline.com/doi/abs/10.1080/03071847.2016.1174477 (accessed 18 March 2021).

CCT Dublin, 2016. The Brexit: what is means for students. Available from: www.cct.ie/the-brexit-what-it-means-for-students/ (accessed 12 December 2020).

Cressey, D., 2016. Brexit by the numbers: the fear of brain drain. *Nature Trend Watch*. Available from: www.nature.com/news/brexit-by-the-numbers-the-fear-of-brain-drain-1.21142 (accessed 13 December 2020).

DCU, 2017. Founding regulation of the Brexit Institute of Dublin City University. Available from: http://dcubrexitinstitute.eu/wp-content/uploads/2017/10/Brexit-Institute-Founding-Regulation.pdf (accessed 18 January 2021).

Department for the Economy, 2012. Graduating to success: a higher education strategy for Northern Ireland. Available from: www.economy-ni.gov.uk/publications/graduating-success-he-strategy (accessed 6 December 2020).

European Commission, 2020a. The EU-UK trade and cooperation agreement. Available from: https://ec.europa.eu/info/relations-united-kingdom/eu-uk-trade-and-cooperation-agreement_en (accessed 1 February 2021).

European Commission, 2020b. Negotiating documents on Article 50 with the United Kingdom. Available from: https://ec.europa.eu/commission/brexitnegotiations/negotiating-documents-article-50-negotiations-united-kingdom_en?field_core_tags_tid_i18n=351 (accessed 3 November 2020).

European Commission, 2020c. Documents related to the work of the Task Force for Relations with the United Kingdom. Available from: https://ec.europa.eu/info/european-union-and-united-kingdom-forging-new-partnership/publications-and-news/documents-related-work-task-force-relations-united-kingdom_en (accessed 3 November 2020).

Gormley-Heenan, C. and Aughey, A., 2017. Northern Ireland and Brexit: three effects on 'the border in the mind'. *British Journal of Politics and International Relations*, 19 (3), 497–511. Available from: https://journals.sagepub.com/doi/abs/10.1177/1369148117711060 (accessed 18 March 2021).

Hay, I., 2016. *Qualitative research methods in human geography*. Oxford: Oxford University Press.

Hayward, K., 2018. The pivotal position of the Irish border in the UK's withdrawal from the European Union. *Space and Polity*, 22 (2), 238–254. Available from: www.tandfonline.com/doi/abs/10.1080/13562576.2018.1505491 (accessed 18 March 2021).

Hayward, K. and Murphy, C., 2018. The EU's influence on the peace process and agreement in Northern Ireland in Light of Brexit. *Ethnopolitics*, 17 (3), 276–291. Available from: www.tandfonline.com/doi/abs/10.1080/17449057.2018.1472426?journalCode=reno20 (accessed 18 March 2021).

Higher Education Authority, 2011. National strategy for higher education 2013. Available from: https://hea.ie/resources/publications/national-strategy-for-higher-education-2030/ (accessed 6 December 2020).

Jakola, F. and Prokkola E-K., 2017. Trust building or vested interest? Social capital processes of cross-border co-operation in the border towns of Tornio and Haparanda. *Tijdschrift voor economische en sociale geografie*, 109 (2), 224–238. Available from: https://onlinelibrary.wiley.com/doi/abs/10.1111/tesg.12279 (accessed 18 March 2021).

Koch, K., 2018a. Geopolitics of cross-border cooperation at the EU's external borders. Doctoral thesis, University of Oulu, Oulu.

Koch, K., 2018b. The spatiality of trust in EU external cross-border cooperation. *European Planning Studies*, 26 (3), 591–610. Available from: www.tandfonline.com/doi/abs/10.1080/09654313.2017.1393502 (accessed 29 March 2021).

Kolossov, V., 2005. Border studies: changing perspective and theoretical approaches. *Geopolitics*, 10 (4), 606–632. Available from: www.tandfonline.com/doi/abs/10.1 080/14650040500318415?journalCode=fgeo20 (accessed 18 March 2021).

Komarova, M., 2017. Cross-border student mobility in third level education: an update. Centre for Cross Border Studies Briefing Paper, October 2017. The Centre for Cross Border Studies. Available from: http://crossborder.ie/site2015/ wp-content/uploads/2017/10/Cross-border-student-mobility-update-paper.pdf (accessed 18 March 2021).

Komarova, M. and Hayward, K., 2019. The Irish border as a European Union frontier: the implications for managing mobility and conflict. *Geopolitics*, 24 (3), 541–564. Available from: www.tandfonline.com/doi/abs/10.1080/14650045. 2018.1496910?journalCode=fgeo20 (accessed 18 March 2021).

Korhonen, J.E., Koskivaara, A., Makkonen, T., Yakusheva, N., and Malkamäki, A., 2021. Resilient cross-border regional innovation systems for sustainability? A systematic review of drivers and constraints. *Innovation: The European Journal of Social Science Research*. Available from www.tandfonline.com/doi/full/10.1080/ 13511610.2020.1867518 (accessed 18 March 2021).

LERU, 2020. Joint statement: EU and UK research and higher education organisations plan a strong future relationship post Brexit. *League of European Research Universities*, 31 January 2020. Available from: www.leru.org/news/eu-and-uk-research-and-higher-education-organisations-plan-a-strong-future-relationship-post-brexit (accessed 13 December 2020).

Malkki, L. and Sinkkonen, T. 2016. Political resilience to terrorism in Europe: introduction to the special issue. *Studies in Conflict & Terrorism*, 39 (4), 281–291. Available from: www.tandfonline.com/doi/abs/10.1080/1057610X.2016.1117 325?journalCode=uter20 (accessed 18 March 2021).

Marque, O. and Miralles-Guasch, C., 2018. Resilient territories and mobility adaptation strategies in times of economic recession: evidence from the metropolitan region of Barcelona, Spain 2004–2012. *European Urban and Regional Studies*, 25 (4), 345–359. Available from: https://journals.sagepub.com/doi/abs/10.1177/ 0969776417703158?journalCode=eura (accessed 18 March 2021).

McCall, C., 2013. European Union cross-border cooperation and conflict amelioration. *Space and Polity*, 17 (2), 197–216. Available from: www.tandfonline.com/ doi/abs/10.1080/13562576.2013.817512 (accessed 18 March 2021).

McCall, C., 2018. Brexit, bordering and bodies on the island of Ireland. *Ethnopolitics*, 17 (3), 292–305. Available from: www.tandfonline.com/doi/abs/10.1080/1744 9057.2018.1472425 (accessed 18 March 2021).

Mitchell, K., 2012. Student mobility and European identity: Erasmus study as a civic experience? *Journal of Contemporary Research*, 8 (4), 490–518. Available from: www.jcer.net/index.php/jcer/article/view/473 (accessed 18 March 2021).

Murphy, M.C. and O'Brennan, J., 2019. Ireland and crisis governance: continuity and change in the shadow of the financial crisis and Brexit. *Irish Political Studies*, 34 (4), 471–489. Available from: www.tandfonline.com/doi/full/10.1080/07907184.2 019.1687621 (accessed 18 March 2018).

O'Keeffe, B. and Creamer, C., 2019. Models of cross-border collaboration in a post-Brexit landscape – insights from external EU borders. *Irish Geography*, 52 (2), 153–173. Available from: http://irishgeography.ie/index.php/irishgeography/ article/view/1399 (accessed 18 March 2021).

Paasi, A., 2012. Border studies reanimated: going beyond the territorial/relational divide. *Environment and Planning A*, 44 (1), 2303–2309. Available from: https://journals.sagepub.com/doi/10.1068/a45282 (accessed 18 March 2021).

Pritchard, R. and Slowey, M., 2017. Resilience: a high price for survival? The impact of austerity on Irish education, South and North. *In*: E. Heffernan, J. McHale and N. Moore-Cherry, eds. *Debating austerity in Ireland: crisis, experience and recovery*. Dublin: Royal Irish Academy, 175–190.

Prokkola, E-K., 2011. Cross-border regionalization, the INTERREG III A initiative, and local cooperation at the Finnish-Swedish border. *Environment and Planning A*, 43, 1190–1208. Available from: https://journals.sagepub.com/doi/10.1068/a43433 (accessed 18 March 2021).

Prokkola, E-K., 2019. Border-regional resilience in EU internal and external border areas in Finland. *European Planning Studies*, 27 (8), 1587–1606. Available from: www.tandfonline.com/doi/abs/10.1080/09654313.2019.1595531?journalCode=ceps20 (accessed 18 March 2021).

Prokkola, E-K., 2022. Resilience thinking and the taken for granted nature of borders. *In*: D.J. Andersen and E-K. Prokkola, eds. *Borderlands resilience – transitions, adaption and resistance at borders*. London: Routledge, 21–36. .

Queen's University Belfast, 2021. Brexit: insight and analysis. *Queen's on Brexit*. Available from: www.qub.ac.uk/brexit/ (accessed 18 January 2021).

RIA, 2016a. Royal Irish Academy Advice Paper on the Future Funding of Higher Education Ireland. Advice Paper No. 8/2016. Royal Irish Academy. Available from: www.ria.ie/sites/default/files/royal-irish-academy-advice-paper-on-the-future-funding-of-higher-education-in-ireland_1.pdf (accessed 18 March 2021).

RIA, 2016b. *The UK and Ireland: the UK's referendum on EU membership: the implications for Northern Ireland's Higher Education Sector*. The Royal Irish Academy. Available from: www.ria.ie/sites/default/files/ni_higher_education-1_0-1.pdf (accessed 18 March 2021).

RIA, 2017a. Royal Irish Academy Brexit Taskforce survey results: impacts and opportunities for higher education and research on the Island of Ireland Post Brexit. Survey Results Analysis no. 1/ 2017. Royal Irish Academy. Available from: www.ria.ie/sites/default/files/ria_brexit_taskforce_survey_results_report_final_0.pdf (accessed 18 March 2021).

RIA, 2017b. *Research and higher education on the Island of Ireland after Brexit*. Royal Irish Academy. Available from: www.ria.ie/sites/default/files/roi_brexit_report-_e-version-1.pdf (accessed 17 November 2020).

RIA, 2020. *RIA Brexit Task Force*. Royal Irish Academy. Available from: www.ria.ie/policy-international/working-groups/ria-brexit-taskforce (accessed 3 November 2020).

Rogers, P., Bohland, J.J., and Lawrence, J., 2020. Resilience and values: global perspectives on the values and worldviews underpinning the resilience concept. *Political Geography*, 83, 102280. Available from: www.sciencedirect.com/science/article/abs/pii/S0962629820303437 (accessed 18 March 2021).

Rumford, C., 2006. Theorizing borders. *European Journal of Social Theory*, 9 (2), 155–169. Available from: https://journals.sagepub.com/doi/10.1177/1368431006063330 (accessed 18 March 2021).

Rumford, C., 2012. Towards a multiperspectival study of borders. *Geopolitics*, 17 (4), 887–902. Available from: www.tandfonline.com/doi/abs/10.1080/14650045.2012.660584 (accessed 18 March 2021).

Salter, M.B., 2013. To make move and let stop: mobility and the assemblage of circulation. *Mobilities*, 8 (1), 7–19. Available from: www.tandfonline.com/doi/abs/10.1080/17450101.2012.747779 (accessed 18 March 2021).

Simon, S. and Randalls, S., 2016. Geography, ontological politics and the resilient future. *Dialogues in Human Geography*, 6 (1), 3–18. Available from: https://journals.sagepub.com/doi/10.1177/2043820615624047

Slusarciuc, M., 2017. Milestones for the resilience of cross-border regions. *CES Working Paper*, 9 (3), 401–422. Available from: https://ideas.repec.org/a/jes/wpaper/y2017v9i3p401-422.html (accessed 18 March 2021).

Tannam, E., 2018. Intergovernmental and cross-border civil service cooperation: the good friday agreement and Brexit. *Ethnopolitics*, 17 (3), 243–262. Available from: www.tandfonline.com/doi/abs/10.1080/17449057.2018.1472422 (accessed 18 March 2021).

TCD, 2020. Collaboration 2013–2018. *Trinity Research*. Available from: www.tcd.ie/research/about/collaborations/ (accessed 12 December 2020).

4 Politics of resilience … politics of borders? In-mobility, insecurity and Schengen "exceptional circumstances" in the time of COVID-19 at the Spanish-Portuguese border[1]

María Lois, Heriberto Cairo and Mariano García de las Heras

Introduction

The imaginations and practices about borders used by the central state and communities in the borderlands have not always matched. This is the case in the Spanish-Portuguese border – called, interestingly, the "line" (*raya/raia*)[2] by the borderland populations. Already Cairo and Godinho (2013) have argued about the contrasting logics present in the border, that is, the state's and the local populations' logics, when the Spanish-Portuguese borderline was demarcated, in 1864 and subsequently. In fact, the main contradiction in the process of demarcation was not between Spanish and Portuguese governments, but between both central governments and the local populations in the borderlands. This is not strange, because the border of a state is related to striated spaces, while the borderlands are smoother spaces (see also Andersen and Prokkola 2021). However, both kinds of spaces are simultaneous, and, as Deleuze and Guattari say,

> What interests us in operations of striation and smoothing are precisely the passages or combinations: how the forces at work within space continually striate it, and how in the course of its striation it develops other forces and emits new smooth spaces.
>
> (1987, p. 500)

The COVID-19 pandemic has caused not only a public health crisis, but also a convulsion of the foundations on which so-called *normality* is based, including that of the borders and borderlands, particularly in the *raya/raia*. Following Ramonet (2020), the pandemic could be categorized as a total social fact, or as an event of our time (Arribas 2020, p. 1), that is, as a process and experience through which institutions, facts, actors, decisions and values are questioned.

DOI: 10.4324/9781003131328-5

Uncertainty, trauma, mistrust and discomfort associated with the collective experience of the situation result from the deployment of exceptional spaces and times destined to mitigate the effects of the pandemic. States of alarm and emergency have rearranged daily life in the form of quarantines, isolation, mobility restrictions and closure of borders. All these measures have a very long historical trajectory and have been used, recurrently, to complement or supplement public health measures in contexts of insecurity, that is, when there is a perceived and identified risk to life in current and unforeseen times.

However, precisely because the measures are the product of a political decision and shape the social construction of exceptionality, it seems important to reflect on how some measures are taken for granted as a means to maintain public security, and, particularly, how such measures are recurrently implemented in what is taken to be "exceptional times" and thereby creative of exceptional spaces or something similar. This is the case of one of them, the closure of borders, that has become in the European context – and especially since the "long summer of migration" of 2015 – one of the common tools with which governments deal with any "exceptional circumstance." In general terms, the closure forms part of the governance of mobility (San Martín Segura 2019) that is fundamental to understand the complex logics of borders, which does not only work to separate but also to connect. In a European Union (EU), which has emphasized the value of internal smooth spaces, the closure of borders is associated with striated spaces, in the sense of Deleuze and Guattari (1987); that is, it is associated with the control of routes between points, showing the resilience of state borders. However, and as we shall see, the smooth spaces of borderlands are able to resist the operation of closure.

In this chapter, we will turn to some of the practices and representations about European borders, external and internal, in times of COVID-19, to approach the resilience of both the state borders and the borderlands. Mobility restrictions associated with the (re)enactment of borders seem to be a persistent state territorial response in times of emergency. However, there is also a re-making and valorization of social relations, practices and representations around the border in borderland communities in times of the closing of borders. Our aim is to contribute to understanding resilience in relation to borders at different scales. We approach the closing of the EU external and internal borders in response to COVID-19 as a border securitization pattern, assuming that the virus would be spread by mobile people coming from abroad more than within the state population. By underlining the changes and definitions accounted for in this closing, especially linked to the Schengen-defined "exceptional circumstances," border state resilience (top-down) will be presented. Secondly, the borderland resilience, that is, local people's experiences, imaginations and adaptation coping with the closure of the border, will be illustrated. Drawing on various materials, documents and fieldwork, we will explore the role of historical memory in borderlands resilience and discuss the interplay of resilience in multilayered and multi-sited border narratives and representations. The reconstruction of borders through institutional and non-institutional actors underline some of the paradoxical border imaginations and practices in times of the pandemic in the context of the European Union. We thereby de-naturalize the border imagination as a security artifact for the EU and the EU states and underline other imagination and practices of the local communities that show the borderlands' resilience.

The empirical focus is on the Spanish-Portuguese border, where the response by both state governments to the in-mobility EU regulations to fight in-security has been a re-making of the internal borders (state scale resilience), while in the borderlands there were developed experiences of cooperation and solidarity (locality scale resilience). The latter illustrates the strength of bonds in border communities and their resilience in order to renew and confront the imagination and reconfiguration of borders after the pandemic.

Border state resilience and borderland resilience

Recent years have witnessed a growing academic consensus that bases the concept of resilience on the idea of adaptation, transformation and survival in contexts of crisis or historical situations affected by adverse circumstances (Gunderson et al. 2002; Bourbeau 2018). The dominant perspectives in the specialized literature consider resilience as an expression of resignation and adaptation as a function of neoliberal governance systems (Evans and Reid 2014; Mavelli 2019) or of the alterations motivated by the changing physiognomy of perceived threats, both real and imaginary (Zebrowski 2008; Michelsen and De Orellana 2019; Wandji 2019). However, as stated in the Introduction (Andersen and Prokkola 2022, p. 6), these arguments ignore the innovative capacity of resilience through creative and transformative responses. This idea connects with the theoretical proposals raised in the renewal of some research agendas in critical border studies recovering the inherently political dimension of borders, referred to as borderscapes and reconceived as sites of generative struggles where alternative subjectivities and agencies could be shaped (Brambilla et al. 2016; Brambilla and Jones 2020; thereto, see also Lamour and Blanchemanche 2022).

The difficulty of its precise definition cannot lead us to interpret resilience as a catch-all word, as "resilience relies on ideas of self-organisation, adaptation, transformation and survival in the face of adversity or crisis" (Humbert and Joseph 2019, p. 215). This makes the perspective of resilience particularly apt for examining the reactions to the pandemic of COVID-19 in relation to borders and in borderlands. Firstly, we underline the plural character manifested by the resilient discursive practices expressed by the multiple socio-political agencies that intervene in border spaces. These actions manifest the articulation of identification processes in constant negotiation at different scales. The research devoted to outlining a basic theory about the border phenomenon (e.g., Newman 2003; Paasi 2011) describes how the theoretical reflections on the drawing of borders advance from classical conceptions that underline their rigidity towards perceptions recognizing the ambivalences of their meanings and, mainly, the processual character of borders. Borders as we know them from modernity grew as part of the states and their creation, and political identifications are built through "bordering practices" (Kuus 2010, pp. 671–672), understood as "a wide range of transformative processes and affective in which social and spatial orders and disorders are constantly reworked" (Woodward and Jones 2005, p. 236). Rethinking the borders through bordering practices implies understanding these practices as something implicit in the construction of those borders, not analyzable as developments incomplete or finished, "but in a constant process of materialization" (Prokkola 2008, p. 15). The idea of resilience allows us to understand how this process is not linear and re-uses past practices, institutions and representations, as we will see later.

A recent fruitful line of research in specialized studies on security is the juxtaposition between the notion of resilience and the border phenomenon (Parizot et al. 2014), in the sense that the tightening of controls in the border leads to advantages for "experts" in crossing borders. These "experts" are borderlanders that used to cross the border for very different reasons (family visiting, farming, shopping. . .) and who, in the new situation, become smugglers, *coyotes* etc. Such approaches support an idea of border based on the recognition of a multiplicity of actors intervening through a wide repertoire of social practices, simultaneously recognizing the political, social and economic connotations underlying the layout of various territorial boundaries. The previous introduces cross-border institutional actions, both formal and informal, in the analysis of the resilience expressed in borderlands (Prokkola 2019).

States (and supra-state organizations, such as the EU) produce borders through the affirmation of images based on specific ideas of protection against threats. The construction of these imaginaries requires the identification of otherness to establish their own "national" identity, which allows assigning security attributes to the border that separates "our" space from "theirs." The borderland communities often dispute these meanings of the borders, and mobilize collective imaginaries that constitute their cultural heritage (often common in both sides of the borders) through different instruments: "This nature of borders leads to the situation that resilience of local communities and the borderland area sometimes has a different meaning than the resilience of the state border as the element of the whole country" (Porczyński and Wojakowski 2020, p. 794). For example, the maintenance of the landmarks that demarcate the border is part of the creation of an official memory on the border, which can be answered by the collective memories of the inhabitants of the border areas, which resist the official memories, through practices such as the demolition and disappearance of landmarks. In the first case, the border will be resilient, while in the second one there is resilience of a non-bounded space that constituted the borderlands.

To understand better the differences between the resilience of the state border and the resilience of the borderlands, we have decided to use a multi-scalar approach to resilience in the context of border closures due to COVID-19. We must consider, at least, three scales of analysis: the macro-regional (EU), the state (Spain and Portugal) and local (the *raya/raia*, the borderlands). The first two allow us to analyze resilience in striated spaces, with clear limits, although the bordering practices in both of them are not always coincident. Merje Kuus points out shrewdly the contradictions between nation-states and macro-regions like the EU, and puts it in relation with resilience: "That even in that most supranationalized space nation-states can effectively veto supranational regulations tells us something about the resiliency of national institutions" (2020, p. 1188). The third scale, the borderlands, allow us to see how resilience works in a smooth space. Borderlands are "smooth" spaces in the sense of Deleuze and Guattari because they are characterized by movement and are fuzzy spaces, without limits. Only the institutionalized borderlands, the euro-regions or euro-cities, have precise limits.

Moreover, examining resilience at the multiple scales implied in the social actions helps to overcome the explanatory limitations constrained by the division established between rupture and continuity in the bordering processes. The multi-scale analysis of resilience is, according to Folke et al. (2010),

> fundamental to understand the interaction between persistence and change. . . . Without the scale dimension, resilience and transformation can seem contradictory and even conflictive. Confusion arises when resilience is interpreted as a look back, assumed to avoid novelty, innovation and transitions towards new development paths.

The study is based on different sets of materials that provide insights into the different versions of the closure of the Spanish-Portuguese border due to COVID-19, and the reactions of borderlanders to the closure. The main documented materials are media reports and social media, which provide us with insight into the practices of security/insecurity that were conducted by state and non-state actors. All these are supported by continuous fieldwork and research in the area for more than 15 years, which alleviates the difficulty of doing fieldwork since the pandemic has exploded. In the following sections, the research material and findings are organized and interpreted to discuss the resilience of the Spanish-Portuguese border and the *raya/raia* as a multiple fact and how the different versions of resilience are played out.

State border resilience in the EU: Tracing a "security perimeter"

States' response to the pandemic brought about general closure of borders all over the world. According to the second Report on Travel Restrictions due to COVID-19 of the World Tourism Organization (UNWTO), issued on April 28, 2020, "100% of destinations now have restrictions in place." Although the measures taken for the different states were not homogeneous, the circumstances were extraordinary: "Never before in history has international travel been restricted in such an extreme manner" (UNWTO 2020, p. 2).

In the context of the European Union, the decision to close the external borders was taken on March 17, 2020. In the weeks and days before, representations of the global space around the association between COVID-19, mobility and security were made by political representatives of the far right such as the Prime Minister of Hungary, Viktor Orbán; the Senator, ex-Minister of Interior and leader of Northern League in Italy, Matteo Salvini; and the Spanish Congressman and General Secretary of Vox, Javier Ortega Smith. All of them were locating the origins, the transmission routes and/or the combat paths for the virus in Europe. As of March 17, 2020, the European Council agreed, for the first time in the history of the institution, to close the external borders of the Union. The closure was initially for 30 days, but open to later extensions. In the official statements, a simultaneous and uniform action on borders was proposed to be applied to the so-called EU+ space, that is, the member states of the Schengen space and the four states associated with the agreement (Iceland, Liechtenstein, Norway and Switzerland) (European Commission 2020). The only mobility allowed would be commodities, and that associated with certain travelers considered essential to ensure the continuity of the common space. In any case, the communication established and sealed a link between borders, mobility and security – in this case, biosecurity – by launching a space where:

> The EU's external border has to act as a security perimeter for *all Schengen States.* It is of common interest and a common responsibility. In the current circumstances, with the coronavirus now widespread throughout the EU, the external border regime offers the opportunity of concerted action among Member States to limit the global spread of the virus. Any action at the external border needs *to be applied at all parts of the EU's external borders.* A temporary travel restriction could only be effective if *decided and implemented by Schengen States for all external borders at the same time and in a uniform manner.*
> (European Commission 2020; highlighted as in the original)

That uniform and coordinated pattern to design a security perimeter, where the closure of borders remains as the main application to restrain mobility by invoking security, would contrast with the unilateral measures of the member states regarding the closure or not of internal borders. This is an example of the already-noticed contradictions between nation-states and macro-regions. As a matter of fact, the Schengen code has always allowed a re-establishment of internal border controls depending on the existence of exceptional circumstances, regulated in Chapter II of the Regulation, amended in 2016 (Regulation UE2016/399), especially in Articles 25–30. In fact, "exceptional circumstances" have ranged from sporting events to visits from the Pope, meetings of the G7, NATO, climate summits, terrorist attacks or meetings of the Hell's Angels (Lois 2014, p. 255; for an intensive study case of Portugal and Lisbon NATO summit 2010, see Amante 2019). In times of COVID-19, the "exceptional circumstances" to reinstate border controls were due to public health issues, something that did not happen in the case of Influenza A or Ebola. However, at least 14 countries (Austria, Czech Republic, Denmark, Hungary, Lithuania, Poland, Germany, Estonia, Portugal, Spain, Finland, Belgium, Switzerland and Norway) rehabilitated their internal borders to show again their resilience as an element in controlling the mobility of people (Lois 2020). As stated earlier, in the cases of Austria, Denmark, France, Germany, Norway and Sweden, controls at some of their borders – although not all of them; for example, Sweden controls only its southern borders, or the northern border to Finland has been open all the time – remained active after the 2015 long summer of migration, and were prolonged even beyond the maximum two years established in Article 29 of the Schengen code (De Miguel 2018). It is precisely this article which launches the security link between external and internal borders, based on a border efficiency necessary to eliminate controls. As stipulated by the European Parliament and the Council:

> In exceptional circumstances where the overall functioning of the area without internal border control is put at risk as a result of persistent serious deficiencies relating to external border control as referred to in Article 21, and insofar as those circumstances constitute a serious threat to public policy or internal security within the area without internal border control or within parts thereof, border control at internal borders may be reintroduced in accordance with paragraph 2 of this Article for a period of up to six months. That period may be prolonged, no more than three times, for a further period of up to six months if the exceptional circumstances persist.
>
> (European Parliament and Council 2016)

Thus, "boring" EU borders – that is, those "open" borders that are still routinely borders: "Cross-border spaces of absence . . . are spaces of boring and bordered everyday practices" (Strüver 2005, p. 219) – become key instruments for the territorialization of a security space, for the preservation of its perimeter. In this case, in the implementation of the closure, the EU internal borders are also projected and concentrated as a security device, with different gradations and identifications of the territorial origins of the danger. In the cases of Denmark, Slovakia or Poland, the closure of the borders would identify the entry of foreigners, in a broad and open sense, like a movement to be controlled. In the case of Greece, the origin of the contagion chains is identified with Spain, Italy, Turkey and the United Kingdom; in the case of the Czech Republic, it is identified with a list of mobilities from EU countries (Belgium, Denmark, Spain, France, Italy or the Netherlands), but also to Switzerland, Norway, UK, China, South Korea and Iran. Closing the circuit of mobility allowed at EU internal borders, Germany, Belgium, Finland or the Netherlands allowed citizens to cross their borders, inside the Schengen space. Again, the association between border closure, control of mobility and identification of legal and healthy mobility abound in the imagination of the border as a territorial device of effective control.

In the case of the Spanish-Portuguese border, and under Article 28 of Schengen, on March 17, 2020 the 64 terrestrial border crossings (OTEP 2017) were closed by agreement between both governments in order "to protect the health and safety of citizens and contain the coronavirus" (Presidencia del Gobierno

Table 4.1 Controls checkpoints at the Spanish-Portuguese border established on March 17, 2020 (from north to south).

Portugal	Spain
Valença*	Tui*
Vila Verde da Raia*	Verín*
Quintanilha*	San Vitero*
Vilar Formoso*	Fuentes de Oñoro*
Termas de Monfortinho	Cilleros
Marvão*	Valencia de Alcántara*
Caia-Elvas*	Badajoz*
Vila Verde de Ficalho	Rosal de la Frontera
Vila Real de San António*	Ayamonte*

* These were "first-class" custom checkpoints in 1960.
Source: www.rayanos.com/noticias/-turismo/551-de-valenca-a-ayamonte-asi-son-los-nueve-pasos-fronterizos-que-resisten-al-covid-19.html and BOE no. 237, 3 October 1960.

2020). Rail connections and airports were also closed to passengers, with a few exceptions. These restrictions did not apply to the transport of goods "in order to ensure the continuity of economic activity and to preserve the supply chain" (Presidencia del Gobierno 2020). The Portuguese prime minister, António Costa, was very explicit when he announced the decision of closure of the border, initially, to leisure trips and tourism: "These measures will be agreed in order to guarantee that our land border will continue to allow the smooth functioning of relations between the two countries, but also the security that is essential to guarantee" (XXII Governo 2020).

However, nine border checkpoints along the *raya/raia* were retained (see Table 4.1).

The governance of mobility at the border went back in time at least 60 years, if not much more. Most of the checkpoints "revived" were the main ones, "first-class custom checkpoints" (*aduanas con habilitación de primera clase*), established by a governmental decree (Jefatura del Estado 1960). Vila Verde de Ficalho–Rosal de la Frontera was a secondary one (*aduanas con habilitación de segunda clase*), and the only new one was Termas de Monfortinho–Cilleros, which was created later. The only other "first-class custom checkpoint" from 1960, Barca d'Alva–Fregeneda, is now in one of the most abandoned sections of the border after the railway between Porto and Spain was closed.

This is proof that the smooth spaces can become striated (and vice versa). "It is a vital concern of every State . . . to control migrations and, more generally, to establish a zone of rights over an entire 'exterior,' over all of the flows traversing the ecumenon," as Deleuze and Guattari (1987, p. 385) state, and the demarcation of the Spanish-Portuguese border in 1864 (and all the associated measures of control) striated the space. While the construction of the European Union (particularly after Schengen) smoothed the internal state borders, allowing the development of more fluid communication in the borderlands – even new institutional forms of smooth space, like cross-border cooperation – it hardened the external ones. Once the state borders are re-closed, "there is still a need for fixed paths in well-defined directions, which restrict speed, regulate circulation, relativize movement, and measure in detail the relative movements of subjects and objects" (Deleuze and Guattari 1987, p. 386). The chosen paths for allowing terrestrial circulation of people, in the case of COVID-19–related closure of the Spanish-Portuguese border, were the former customs checkpoints, proving their resilience.

Resilience as resistance in the borderlands

COVID-19 institutional restrictions on mobility and border crossing have revealed another resilience, that of the *raya/raia* – that is, the borderland. On the one hand, borderlanders seem to turn to what were traditional, everyday or occasional relations between populations on either side of the border. On the other hand, a more recent trend, seem to be border practices and projects where the borderland become a space of possibility, a place of meetings and a place of reference in the construction of political communities. As previously stated, in the borderlands the limit is the center, a reference to the opening of diverse spatialities: "The border can be a material and symbolic resource for local populations, who reinvent it, transgress it, saturate it, or reproduce it in their daily lives, tracing variable territorialities, more complex and unfinished around it" (Lois 2017, p. 95). Hence, the gaze of the local community can be seen to confront the gaze of both the EU and the State on the border. This encounter shows the scope of borderland thinking and practice trapped in another territorial logic, in the territorial trap (Agnew 2003) of the borderline understood as delimitation of security spaces, as its confines.

However, seen from the borderland, the closing of the Spanish-Portuguese border sends a contrasting signal, thus revealing multiple and creative bordering possibilities. In the Miño/Minho area, in the northern part of the borderland, local practices and imaginations confront international and global spaces' times and practices. For the mayor of Tui (Spain), one of the busiest crossings along this border, restoration of border controls over an international bridge that links the Spanish and Portuguese borderlands brings back memories of the border in past times. In his own words, "cutting the bridge is like building the Berlin wall" (Jiménez 2020) – because the development of communication infrastructures has been very important in the (re)construction of borderland practices and imaginations.

In the same area, in Salvaterra (Spain), mayor Marta Valcárcel stated:

> This closing is making a border like it never was. Before the bridge there was the ferry, and previously small boats, and the river was even crossed by swimming. This is the first time, as far as I remember, that traffic is cut off, day-to-day. We are not aware that we go from one country to another, because in the end we are one. This is something totally new.
>
> (Martínez and Rodríguez 2020)

In an explicit reference to the limitation of having to cross the border at the border post, the mayor herself refers to the difficulty of cross-border workers, now mediated by a detour of about 60 kilometers (Martínez and Rodríguez 2020). In this case, border closing introduces a scenario of the struggle of borderlanders against central governments. Uxío Benítez, director of the "Río Miño/Minho" European Grouping of Territorial Cooperation (EGTC), when announcing the call to a demonstration by the local authorities of the municipalities of Northern Portugal and Southern Galician members of the EGTC (Tomiño, Tui, Salvaterra, As Neves, Arbo, A Guarda, O Rosal, Melgaço, Valença, Vila Nova de Cerveira, Paredes de Coura y Monçao), stated it clearly: "the economic reality of the *raia* is unknown in Madrid and Lisbon" (Garrido 2020). It was a significant demonstration that took place on the Bridge of Friendship (*Puente de la Amistad*), which links Goián-Tomiño, Galicia, with Vila Nova de Cerveira, Portugal, on June 3, 2020 (EP 2020). The local authorities demanded the re-opening of all crossing passes, constructing the word "SOS" with big letters in the middle of the international bridge, because they considered that open borders give opportunities – and, in the last instance, can become the economic solution – for borderland working communities, already depressed by long years of marginality.

The Portuguese maritime police were reporting signs of movements and furtive activities in the Miño/Minho river; in particular, in Monção and Melgaço areas, on the Portuguese side. The police had found artisanal devices (chords moved by pulleys that allow the exchange of products between both shores) to

Figure 4.1 Demonstration by the local authorities of the municipalities of Northern Portugal and Southern Galicia on the Bridge of Friendship (Puente de la Amistad), which links Goián-Tomiño.

Source: Photo by J. Lores. Rights given up by *Faro de Vigo*.

break the confinement between both countries (Fernandes 2020). In the Alto Miño/Minho, where there is a land border, borderland people were aware of movement of persons across the border using old paths, trails and roads now closed by central governments due to the pandemic. They were recovering, on the one hand, the activities (exchange of products) typical of smuggling as the common strategy to cope with restrictions in times of dictatorships on both sides and, on the other hand, the movements and itineraries for smuggling of both commodities and people to migrate to France (Fernandes 2020). The creative solutions already employed historically in hard times, when personal survival was the foremost concern, are reconfigured in certain recognized socio-spatial ways, showing their resilience.

Following the borderline, a little further south, we find Rio de Onor (Portugal) and Rihonor de Castilla (Spain), which is a binational village – an exception to the national exclusionary imaginations celebrated in the Treaty of Limits of 1864. After the first closure, the Spanish and Portuguese governments agreed to allow the opening of the border a few hours a week, on Tuesdays and Saturdays. The displacements associated with the agricultural and livestock jobs follow itineraries where the border crossing is constant, showing the invisibility of the border for daily materiality. On a more symbolic horizon, locals remember that not even in the times of the Iberian dictatorships did the border close in this way (Rádio e Televisão de Portugal 2020). Curiously, they do know where the territorial limit lies. However, it is that very idea that blurs its existence: "There was never a border here. Here it is Portugal, here it is Spain, but this is only one village" (González Velasco 2020a). The material ties, kinship and conviviality between both sides of the community are practiced in a smooth space superimposed on the linear territoriality of the state border, which is part of a striated space. The state border is surpassed and re-appropriated by borderland resilience practiced in everyday life (thereto, see also Ferdoush 2022; Andersen 2022).

Again, further south, on the border between Fuentes de Oñoro (Spain) and Vilar Formoso (Portugal), despite the border closure, the delivery service of general food, medicines and bread has maintained its daily cross-border itinerary (González Velasco 2020b). In a borderland rural area, which shares demographic emptying, an aging population and scarce services, the maintenance of the cross-border daily activities becomes precisely what allows the locals to resist and to cope with the supervened "exceptional circumstances." It is again the resilience of borderland commonalities that allows local people to overcome the closure of the border.

It is important to underline that we are talking of an imaginary and practices of the population of the borderland, which are not shared by the rest of the Portuguese and Spanish population. This is shown in a media report about the possibility of a new closure of the border after the summer, when the borderlanders were definitely against the closure, but the rest of those interviewed, from the rest of Portugal, would understand a new closure (Lusa 2020). Then, the borderland resilience is not only mediated in bottom-up local agency, but also linked to the everyday making of a specific space, hard to delimit, where the people belong to an imagined community interconnected by their perceived proximity to the border.

Borderland resilience in all these cases shows the adaptability of traditional practices and imaginaries of cohabitation to new circumstances. However, it also shows the less traditional interaction between the two sides of the border developed after the "disappearance" of borders and the extension of the institutional cross-border cooperation. In relation to the closure of the border due to COVID-19, borderland resilience is associated with the resistance of *rayanos/arraianos* to the decisions of central governments, which are perceived as bad and harmful for them (see also Andersen 2020, 169–171).

Conclusions

We have analyzed the resilience of the state borders and of the borderlands in relation to the closing of borders in the context of the COVID-19 pandemic, as a measure to face the so-called "exceptional circumstances." We have specifically researched the Spanish-Portuguese border and borderland, the *raya/raia*. The "exceptional circumstances" shaped in 2020 by a global pandemic bring about the geopolitical imagination and practice of the states that perform the border as a territorial device to control mobility and pledge immobility, and as a tool applicable to the control of the insecurity generated by the pandemic as well.

The treatment of European borders, internal and external, boring and spectacular, is linked in EU regulations, from 2015 but also in COVID-19 times. Hegemonic institutional visions project the ability of borders to manage an effective control of mobility in a delimited perimeter, extended to all its forms, and even recurring to past control mechanisms if it is necessary. In fact, Schengen can be envisaged as a "resilient project" (Votoupalova 2020), and the closure of state borders (internal EU borders) is emphasized as an effective (bio)securitization attribute in institutional discourses and practices during the pandemic.

However, in the borderland communities of the Spanish-Portuguese border, it seems that there are different – and sometimes contradictory – ways in which it is possible to imagine and practice the border. This is revealed in the constant negotiation of the meaning of borders for everyday life, related to memories and experience in the borderland. In the examples we have exposed, daily borderland survival strategies confront the central governments' institutional imagination and practice; whereas the state constructs the image of a "protective" border, the borderlanders imagine and practice the border as "enabling." Borderland resilience sets a border as a common good, very far from the implications of "seeing as a state" as discussed by James Scott (1998). In sum, the different horizons beyond the centrality of state scale, of the territorial view that naturalizes borders as insecurity controls, empirically illustrate and underline different border-related resiliencies.

The multi-scalar research made has shown that there are different kinds of resilience according to the character of the space, smooth or striated. In a nutshell, resilience of state borders and resilience of borderlands are not equivalent (thereto also Frandsen 2022). As it was mentioned at the beginning of this chapter, the imaginations and practices surrounding the Spanish-Portuguese borders, both of the states and of the communities in the borderlands, do not always match. It is not our aim to demonize the state border control to clearly separate one nation from the other and securitize its territory, or idealize the communitarian experiences of the local people of the borderlands, at all; but it is true that the historical memories and practices of control (by the state) and resistance (in the borderlands) have been part of this border landscape for many years. May this be an inspiring path to emancipate the hegemonic gaze on borders, underlining the experiences, memories and creative resistance (and resilience) that historically also conform and shape the character of the borderlands.

Notes

1 This research has been supported by the Research Project Santander – Universidad Complutense de Madrid PR87/19–22689 "Consolidación y difusión de la cooperación transfronteriza: transformaciones de la gobernanza territorial en la Península Ibérica y América Latina."
2 *La raya* (in Spanish) or *a raia* (in Portuguese and Galician) means *line* (in English), and it is used by the borderland populations to refer to the states' political border (*la frontera*, in Spanish, or *a fronteira*, in Portuguese). But it also alludes to the borderland and to the borderlanders; thus, there are *rayanas* (in Spanish) or *arraianas* (in Portuguese and Galician) populations.

References

Agnew, J., 2003. *Geopolitics: re-visioning world politics*. London: Routledge.

Amante, M.F., 2019. Performing borders: exceptions, security and symbolism in Portuguese borders control. *Journal of Borderlands Studies*, 34 (1), 17–30. DOI: 10.1080/08865655.2016.1270169.

Andersen, D.J. 2022. Line-practice as resilience strategy: the Istrian experience. *In*: D.J. Andersen and E-K. Prokkola, eds. *Borderlands resilience: transitions, adaptation and resistance at borders*. London: Routledge, 166–181.

Andersen, D.J. and Prokkola, E-K., 2022. Introduction: embedding borderlands resilience. *In*: D.J. Andersen and E-K. Prokkola, eds. *Borderlands resilience: transitions, adaptation, and resistance at borders*. London: Routledge, 1–18.

Arribas, B., 2020. El acontecimiento de nuestro tiempo: algunas lecciones éticas. *Hercritia, Cátedra Internacional de Hermenéutica Crítica*. Available from: www.catedradehermeneutica.org/pandemia-globalizacion-y-ecologia-03/?fbclid=IwAR30aJy3v9SQEuXuCA4jxQL4jmc9MqHF-i6rthly42Vrv35_3B8PdSvXC48 (accessed 18 November 2020).

Bourbeau, P., 2018. *On resilience: genealogy, logics and world politics*. Cambridge: Cambridge University Press.

Brambilla, C. and Jones, R., 2020. Rethinking borders, violence, and conflict: from sovereign power to borderscapes as sites of struggles. *Environment and Planning D – Society & Space*, 38 (2), 287–305.

Brambilla, C., Laine, J., Scott, J.W., and Bocchi, G., eds., 2016. *Borderscaping: imaginations and practices of border making.* New York: Routledge.

Cairo, H. and Godinho, P., 2013. El Tratado de Lisboa de 1864: la demarcación de la frontera y las identificaciones nacionales. *Historia y Política* (30), 23–54.

De Miguel, B., 2018. Los controles fronterizos dentro de Schengen se perpetúan a pesar de las quejas de Bruselas. *El País*, 12 November. Available from: https://elpais.com/internacional/2018/11/12/actualidad/1542042593_317284.html (accessed 20 November 2020).

Deleuze, G. and Guattari, F., 1987. *A thousand plateaus: capitalism and schizophrenia* (trans. B. Massumi). London: Athlone.

EP, 2020. Alcaldes de la "raia" exigen la apertura de la frontera "al menos para trabajadores". Urgen habilitar más pasos que el de Tui porque este puente "no da para más". *El Faro de Vigo*, 3 June. Available from: www.farodevigo.es/comarcas/2020/06/03/alcaldes-raia-exigen-apertura-frontera-15179199.html (accessed 12 June 2020).

European Commission, 2020. *Communication from the Commission to the European Parliament, the European Council and the Council* (COM 2020, 115 final). Available from: https://ec.europa.eu/transparency/regdoc/rep/1/2020/EN/COM-2020-115-F1-EN-MAIN-PART-1.PDF (accessed 24 November 2020).

European Parliament and Council, 2016. *Regulation (EU) 2016/399 of the European Parliament and of the Council, of 9 March 2016, on a Union Code on the rules governing the movement of persons across borders (Schengen Borders Code).* Available from: https://eur-lex.europa.eu/legal-content/EN/TXT/PDF/?uri=CELEX:32016R0399&from=ES (accessed 10 November 2020).

Evans, B. and Reid, J., 2014. *Resilient life: the art of living dangerously.* Cambridge: Polity Press.

Ferdoush, M.A., 2022. Stateless' yet resilient: refusal, disruption and movement along the border of Bangladesh and India. *In*: D.J. Andersen and E-K. Prokkola, eds. *Borderlands resilience: transitions, adaptation, and resistance at borders.* London: Routledge, 106–118.

Fernandes, A.P., 2020. Pandemia ressuscita antigas rotas clandestinas da emigração a salto e contraband. *Jornal de Notícias*, 26 April. Available from: www.jn.pt/nacional/pandemia-ressuscita-antigas-rotas-clandestinas-da-emigracao-a-salto-e-contrabandistas-12119949.html?fbclid=IwAR0tMZGbQ4yrRAaiMOQ22We0-XAhSuxM5CtOdmQwAFx3LkuTg4Q-UVRTnqk (accessed 2 May 2020).

Folke, C., Carpenter, S.R., Walker, B., Scheffer, M., Chapin, T., and Rockström, J., 2010. Resilience thinking: integrating resilience, adaptability and transformability. *Ecology and Society*, 15 (4), art. 20. Available from: www.ecologyandsociety.org/vol15/iss4/art20/ (accessed 15 June 2020).

Garrido, M., 2020. A raia pontevedresa pasa á mobilización simbólica pola reapertura fronteiriza? *El Correo Gallego*, 3 June. Available from: www.elcorreogallego.es/primer-plano/a-raia-pontevedresa-pasa-a-mobilizacion-simbolica-pola-reapertura-fronteiriza-XK3279624 (accessed 10 June 2020).

González Velasco, P., 2020a. La solidaridad, una "vacuna" contra el cierre de la frontera en La Raya. *El Trapezio*, 29 April. Available from: https://eltrapezio.eu/es/espana/la-solidaridad-una-vacuna-contra-el-coronavirus-en-la-raya_8665.html?fbclid=IwAR1s_wYZ37wXl9J_hsceahXReHHR17-cFQecBjy_y_pQ_A7viKic5cIiX3Y (accessed 2 May 2020).

González Velasco, P., 2020b. Aqui nunca houve fronteira. Aqui é Portugal, aqui é Espanha, mas isto é um povo. *El Trapezio*, 7 May. Available from: https://eltrapezio.eu/pt-pt/espanha/aqui-nunca-houve-fronteira-aqui-e-portugal-aqui-e-espanha-mas-isto-e-um-po-vo_8983.html?fbclid=IwAR0zlomX0TPw8kT1Cliu1v A5FWfx9NR0ItFtQFgplXo9_srnvHTMT6LthOo (accessed 15 May 2020).

Gunderson, L., Holling, C.S., Pritchard, L., and Peterson, G.D., 2002. Resilience. *In*: H.A. Mooney and J.G. Canadell, eds. *Encyclopaedia of global environmental change. Vol 2: the Earth system. Biological and ecological dimensions of global environmental change*. Chichester: Wiley, 530–531.

Humbert, C. and Joseph, C., 2019. Introduction: the politics of resilience: problematizing current approaches. *Resilience: International Policies, Practices and Discourses*, 7 (3), 215–223.

Jefatura del Estado, 1960. Instrumento de ratificación del Convenio Aduanero entre España y Portugal relativo al Tráfico Internacional por Carretera, Ferrocarril y Ríos Limítrofes. *Boletín Oficial del Estado* (BOE), (237), 3 October. Available from: www.boe.es/diario_boe/txt.php?id=BOE-A-1960-14062 (accessed 7 September 2020).

Jiménez, J.L., 2020. Municipios fronterizos: "Cortarnos el puente es como levantar el muro de Berlín". *ABC*, 4 May. Available from: www.abc.es/sociedad/abci-municipios-fronterizos-cortarnos-puente-como-levantar-muro-berlin-202005040121_noticia.html (accessed 10 June 2020).

Kuus, M., 2010. Critical geopolitics. *In*: R. Denemark, ed., *The international studies encyclopedia*, vol. II. Chichester: Wiley-Blackwell, 683–701.

Kuus, M., 2020. Political geography III: bounding the international. *Progress in Human Geography*, 44 (6), 1185–1193.

Lamour, C. and Blanchemanche, P., 2022. A resilient Bel Paese? investigating an Italian diasporic translocality between France and Luxembourg. *In*: D.J. Andersen and E-K. Prokkola, eds. *Borderlands resilience: transitions, adaptation, and resistance at borders*. London: Routledge, 152-165.

Lois, M., 2014. Apuntes sobre los márgenes: fronteras, fronterizaciones, órdenes socioterritoriales. *In*: E. Cardin and A.S. Colognese, orgs., *As Ciências Sociais nas fronteiras. Teorias e metodologias de pesquisa*. Cascavel: Editora JB, 239–259.

Lois, M., 2017. Geopolítica de la Paz y Estudios de Frontera. *La Migraña*, (22), 92–97.

Lois, M., 2020. Los Estados cierran sus territorios por seguridad . . . pero los virus están emancipados de las fronteras. *Geopolítica(s): Revista de Estudios sobre Espacio y Poder*, 11, 293–302. DOI: 10.5209/geop.69370.

Lusa, 2020. Fronteiras fechadas? Portugueses e espanhóis alertam para "suicídio económico". *SIC Noticias*, 8 October. Available from: https://sicnoticias.pt/especiais/coronavirus/2020-10-08-Fronteiras-fechadas–Portugueses-e-espanhois-alertam-para-suicidio-economico (accessed 18 November 2020).

Martínez, R. and Rodríguez, L., 2020. El suspiro de la "Raia" por una desescalada simétrica de España y Portugal. *El Correo Gallego*, 3 May. Available from: www.elcorreogallego.es/primer-plano/el-suspiro-de-la-raia-por-una-desescalada-simetrica-de-espana-y-portugal-EG2796789 (accessed 7 May 2020).

Mavelli, L., 2019. Resilience beyond neoliberalism? Mystique of complexity, financial crises, and the reproduction of neoliberal life. *Resilience: International Policies, Practices and Discourses*, 7 (3), 224–239.

Michelsen, N. and De Orellana, P., 2019. Discourses of resilience in the US alt-right. *Resilience: International Policies, Practices and Discourses*, 7 (3), 271–287.

Newman, D., 2003. On borders and power: a theoretical framework. *Journal of Borderlands Studies*, 18 (1), 13–25.

OTEP, 2017. *Observatorio transfronterizo España-Portugal* (Documento no. 8). Secretaría General de Transporte, Ministerio de Fomento, Madrid. Available from: www.fomento.gob.es/NR/rdonlyres/8D19946C-814B-468D-BBE5-07BAA3 64DC81/143436/Informe_OTEP8.pdf (accessed 27 December 2020).

Paasi, A., 2011. A border theory: an unattainable dream or a realistic aim for border scholars? *In*: D. Wastl-Walter, ed., *The Ashgate research companion to border studies*. Farnham: Ashgate, 11–31.

Parizot, C., Laure, A., Szary, A., Popescu, G., Arvers, I., Cantens, T., and Cristofol, J., 2014. The antiatlas of borders: a manifest. *Journal of Borderlands Studies*, 29 (4), 503–512.

Porczyński, D. and Wojakowski, D., 2020. Borderlands from the resilience perspective: diversification of state borders in former Austrian Galicia. *Regional Science Policy and Practice*, 12 (5), 793–813.

Presidencia del Gobierno, 2020. Restablecidos los controles en las fronteras terrestres en el marco de las medidas de contención del COVID-19. *La Moncloa*, 16 March. Available from: www.lamoncloa.gob.es/serviciosdeprensa/notasprensa/interior/Paginas/2020/160320-covid-fronteras.aspx (accessed 8 November 2020).

Prokkola, E., 2008. *Making bridges, removing barriers: cross-border cooperation and identity at the Finnish-Swedish border*. Oulu: Nordia Geographical Publications.

Prokkola, E., 2019. Border-regional resilience in EU internal and external border areas in Finland. *European Planning Studies*, 27 (8), 1587–1606.

Rádio e Televisão de Portugal, 2020. Parcialmente aberta a fronteira entre Rio de Onor e Rihonor de Castilla. *RTP*, 30 April. Available from: www.rtp.pt/noticias/pais/parcialmente-aberta-a-fronteira-entre-rio-de-onor-e-rihonor-de-castilla_v1225040 (accessed 15 May 2020).

Ramonet, I., 2020. La pandemia y el sistema-mundo. *Le Monde Diplomatique*, 25 April. Available from: https://mondiplo.com/la-pandemia-y-el-sistema-mundo (accessed 3 May 2020).

San Martín Segura, D., 2019. Las lógicas de gobierno de lo fronterizo en el espacio Schengen: la frontera como estriación. *Revista CIDOB d'Afers Internacionals* (122), 15–37. DOI: 10.24241/rcai.2019.122.2.15.

Scott, J.C., 1998. *Seeing like a state: how certain schemes to improve the human condition have failed*. New Haven, CT: Yale University Press.

Strüver, A., 2005. Bor(der)ing stories: spaces of absence along the Dutch-German border. *In*: H. Van Houtum, O. Kramsch and W. Zierhofer, eds. *B/ordering space*. Aldershot: Ashgate, 221–237.

UNWTO, 2020. COVID-19 related travel restrictions: a global review for tourism (second report as of 28 April 2020). *World Tourism Organization (UNWTO)*. Available from: https://webunwto.s3.eu-west-1.amazonaws.com/s3fs-public/2020-04/TravelRestrictions%20-%2028%20April.pdf (accessed 25 December 2020).

Votoupalova, M., 2020. Schengen cooperation: what scholars make of it. *Journal of Borderlands Studies*, 35 (3), 403–423.

Wandji, G., 2019. Rethinking the time and space of resilience beyond the west: an example of the post-colonial border. *Resilience: International Policies, Practices and Discourses*, 7 (3), 288–303.

Woodward, K. and Jones, J.P., 2005. On the border with Deleuze and Guattari. *In*: H. van Houtum, O. Kramsch and W. Zierhofer, eds. *B/ordering space*. Aldershot: Ashgate, 234–248.

XXII Governo, 2020. Fronteira com Espanha fechada para lazer e Turismo. Declaração do Primeiro-Ministro sobre a situação do coronavírus/Covid-19. *República Portuguesa*, 15 March. Available from: www.portugal.gov.pt/pt/gc22/comunicacao/noticia?i=fronteira-com-espanha-fechada-para-lazer-e-turismo (accessed 12 September 2020).

Zebrowski, C., 2008. Governing the network society: a biopolitical critique of resilience. *Political Perspectives*, 3 (1), 1–41.

Part 2

Tracing space

Social relations and movement
as resilience

5 Resilience at Hungary's borders

Between everyday adaptations and political resistance

Sara Svensson and Péter Balogh

Introduction

Hungarian borderlands make up an interesting case for studying resilience in borderlands for several reasons. On the one hand – and similarly to most countries of the world – many of Hungary's borderlands are in a disadvantaged position compared to the country's geographic centers, with the immediate borderlands hosting few major agglomerations or growth hotspots. On the other hand, while socio-spatial disparities are average in a European context, the country overall belongs to the economically less developed member states of the European Union (EU). This means that its citizens, especially in the borderlands, have a particularly strong incentive to take advantage of opportunities offered by various border locations (Sohn 2014).

Hungary belongs to the countries with one of the highest numbers of neighboring states, including five EU (Austria, Slovakia, Romania, Croatia and Slovenia) and two non-EU members (Ukraine and Serbia), and the strongly diverging characteristics and levels of development of Hungary's neighbors imply that challenges and opportunities can be very different in its border regions. In addition, due to massive border shifts in the first half of the 20th century agglomerations formerly belonging to Hungary can be found on the opposite side of its current borders. The issue of ethnic Hungarian minorities has had consistent high salience in Hungarian politics, making borderlands and their populations central to political discourse, while at the same time staying peripheral in terms of economic and social power. While this does add an ethno-linguistic element to cross-border contacts, these are – as will be shown – far from being limited to mono-ethnic encounters.

The chapter starts with a theoretical section, articulating the chapter's approach to resilience and showing how resilience approaches have been applied in research on Hungary. This is followed by three sections, each covering a different process that can also be interpreted as a case of resilience. Three processes are covered: the process of including "the other side" in what it means to be "local," a case of cross-border food communities; the process of securing livelihoods in growing agglomerations and neglected peripheries, a case of cross-border commuting; and the process of resisting the rhetoric and practices of closed borders and societies – a case of cross-border solidarity and humanitarianism.

DOI: 10.4324/9781003131328-7

The case selection allows for a breadth of investigation into activities that are usually not analyzed together yet bound by their joint relation to the border. Solidarity and humanitarianism are not often associated with borderlands in the literature on cross-border cooperation; in the international literature on migration, borders are frequently associated with the cruelty of modern nation-state regimes (Jones 2016) – in the European context, also a supranational regime in the form of the European Union. However, in this chapter we focus on the relation of resilience to resistance among inhabitants of the borderlands. An even more understudied topic in relation to borders is how the production and consumption of food may link people in territories spanning borders and whether that indicates resilience, and if so, what types. Cross-border commuting, on the other hand, is a common theme in border studies, especially from economic and sociological perspectives, but in the analysis of the Hungarian case we see the extent to which the resilience concept is analytically fruitful. With the analysis of these cases, we contribute to the volume's inquiry into the role of borders in social resilience processes and how borderland communities adapt, renew and resist in parts of the world repeatedly experiencing crises, transitions and changing border environments.

Theoretical perspectives on resilience in international and local scholarship

Hungarian borderlands have been subjects to major crises and disruptions such as wars, border revisions and regime change in the 20th century, and they have felt the impact of 21st-century crises such as financial turmoil and the politicization of migration. However, in line with the approach of this volume, we also recognize the presence of slow-moving threats and events, conceptualized as "multiplicity of disruption." Resilience of "borderland people," thus, is situated in diverse contexts and can occur in response to such multiplicity of disruption (Wandji 2019). It consists of "different social groups' ability to self-organize and mobilize skills and resources to create opportunities when faced with adversity and to act in solidarity when their community is disturbed and even disrupted" (Andersen and Prokkola 2022, p. 7). Resilience also goes argue beyond mere coping and adaptation. (As Andersen and Prokkola (Ibid.) note . . .) resilience necessarily locates in a paradoxical relation between adaptation and resistance to change, involving creative meddling with and local interpretations of the changes imposed from 'outside or 'above' (p. 7ff.). However, the concept's relation to resistance is ambiguous as not all resistance to change is resilience. It is important to recognize the political component of resilience, easily overlooked in a field dominated by economic approaches. Our approach therefore opens up for interpretations of resilience as political resistance without equating resilience with resistance.

Two further theoretical perspectives on resilience may be relevant for analyses of processes in borderlands in the present chapter. The first concerns time and distinguishes between "bouncing-back capabilities" and "bouncing-forward capabilities." "Bouncing back" is a somewhat traditional approach and refers to the capacity of communities to maintain structures after instances of crisis or long-term changes. It may also be characterized as robustness. "Bouncing-forward" capabilities, on the other hand, indicate flexibility, the ability to adapt to new circumstances and the capacity to move forward (Shaw 2012), and is, we argue, most in line with the definition given previously. The second approach makes an ontological distinction between resilience as a process, which can only be assessed in hindsight, or resilience as a "stock of something" that can be measured at any given point. The resilience of a community, town or nation is then something that can be given a value and differ from year to year. The latter understanding has undoubtedly entered the competition-oriented logic of modern regional development and led to efforts to develop indices and other yardsticks on resilience (Sensier et al. 2016; Fazekas et al. 2017). In this chapter, we take the process-oriented approach, with the methodological implication that we look for manifestations of resilience within developing timeframes, rather than indicators of resilience produced for the purpose of measurement. It has been shown that the latter will always remain imperfect and thus subject to change (Diaz-Sarachaga and Jato-Espino 2019). At the same time, we acknowledge some methodological limitations. For the analysis, we rely partly on reinterpretation of interviews and documents collected for research that we previously carried out within the context of previous studies,[1] and partly on secondary literature or sources on the Hungarian context.

While the popularity of the concept of resilience among international scholars and policy practitioners has led to several adaptations in the Hungarian context, much of it uses the approach to resilience as a stock of something rather than resilience. For instance, Benke et al. (2018) link the neoliberal emphasis on competition and innovation to a model of learning that can also be used to predict resilience. Notably, their index of learning across Hungarian local governments paints a mixed picture of the situation at Hungarian borders. While there are larger stretches of high-performing communities, for instance at the northwestern border and concentrated around Szeged in the south, there are also small pockets of high-performers in terms of learning capacity embedded even in low-performing border regions. A likewise stock-oriented approach measuring resilience in regions during and shortly after the financial crisis of 2008 found that non-metropolitan regions were more resilient in Hungary (Davies 2011), indicating an economic "bouncing-back" capacity. We identified one study that took the same neoliberal stock-oriented approach to specifically study cross-border areas. Fazekas et al. (2017) proposed and tested a regional homogeneity index on the Western Trans-danubia and Burgenland cross-border region. Finally, a study that differs from others and is interesting in relation to our second case study is Lendvay's (2016) study of a watermelon-producing community (Medgyesegyháza, Hungary), where he advanced the use of resilience and tested its usability for rural communities. In Hungarian, the foreign-sounding and directly translated term *rezilencia* has been applied together with the term *rugalmasság*, better translated as "flexibility." For instance, Pirisi (2017) discusses the role of resilience in the development of small towns, and there is an abundance of semi-academic practitioner reports, often linked in some way to EU funding and/or influence mechanisms. Such semantic distinctions are not unique to the concept of resilience, but serve as a reminder that citizens and actors of borderlands may interpret their own actions differently, perhaps precisely because of such distinctions.

In the empirical sections that follow, we aim at taking a process-oriented approach, looking for resilience in borderlands in response to different types of disruptions – both unexpected major-impact events (e.g., regime change in 1990, refugee spike in 2015 and COVID-19 in 2020) and long-evolving situations (e.g. social, economic and cultural consequences partly related to these crises). Furthermore, "resilience" denotes processes that go beyond coping, adaptation and flexibility, and may move into territories of politically laden resistance revolving around certain values. With this, we recognize recent work linking values and resilience (Rogers et al. 2020). We also aim to see whether the distinction between bouncing-back and bouncing-forward processes is meaningful in the Hungarian context.

Cross-border food communities: including "the other side" in what it means to be "local"

Locally produced and consumed food is often seen as an important component of environmental and economic movements, resisting climate change and neoliberal economic arrangements. At the same time, food production has also become a field of important nationalist symbolism, where countries find it more important encouraging their citizens to buy products produced on their own territory than considering the distance they have been transported. In many European cross-border regions, support of locally produced food has also become a symbolically important area of activity. So far, however, there is little research on what locally produced food means in a cross-border context. Thus, using the example of a shopping community in Esztergom at the Hungarian-Slovak border (Svensson et al. 2019), the first of our empirical cases serves to investigate the extent to which food produced and consumed within borderlands can serve as a material and symbolic source of resilience, and whether local cross-border food communities imply resistance to, or endorsement of, the nationalist discourses surrounding food production.

When coming to power in 2010, Hungary's national-conservative government introduced and developed policies that support local small-scale farming and promote a nationalist discourse around this, manifested by slogans such as "buy Hungarian!" (Lendvai and Ördög 2018, p. 40). At the same time, the overall output of the country's important agricultural sector is heavily skewed towards large-scale units. Farming activities in cross-border regions do not always easily fit these narratives and policies, as we elaborated in a case study of a cross-border food community located an hour's drive northwest of Budapest at the Slovakian border (Svensson et al. 2019).

The Small Basket Shopping community *Kiskosár* was founded in 2011 by a local civil society activist from an environmental movement and located in an area border region comprising the Hungarian town Esztergom and the Slovak town

Štúrovo and their surrounding rural hinterlands. This area has less than 70,000 inhabitants but is known for its intensive cross-border activities (Balogh and Pete 2018; Svensson and Nordlund 2015). In economic terms, the Hungarian side of the border has been somewhat richer, exacerbated by foreign direct investment creating employment opportunities at factories and cross-border commuting, while the Slovak side has high unemployment, out-migration and few industrial workplaces (Balogh and Pete 2018). Due to a large Hungarian-speaking minority on the Slovak side, language difficulties are generally less prevalent than an outsider may assume.

The Small Basket Community is the only shopping community of importance in the region around Esztergom and serves up to a hundred small municipalities around the Danube, Ipoly and Hron rivers. Most in its group of 30 producers are located within 40 kilometers, as is the bulk of its customers. During the build-up phase, they actively sought farmers on both sides of the border to deliver to the community, leading to a peak of a third of the producers being residents of the Slovak side. However, as it became clear that only Slovaks with a registered company in Hungary were allowed to sell in Hungary, the number dropped drastically to less than 10 percent in 2020, something the organization has tried to solve by turning to cross-border cooperation organizations for help to gain attention from decision-makers.

The organization projects an image of itself as part of a larger transnational shopping community movement, seeking to create and sustain personal relationships between producers and consumers. They want to be distinct from the weekly "regular" local product market taking place in the town of Esztergom on Sundays, also rejecting the model in which buyers receive a box of pre-ordered item without face-to-face contact with a producer. In 2020, they had nearly 2,000 regular buyers. Before the 2020 pandemic, these would normally come from both sides of the border and be organized around small-scale bi-weekly collection meetings (2.5 hours long) where buyers "stand in a queue, talk to each other, and become acquainted with each other" (Interview 2016a), a preferred model, able to build a sense of group-belonging, i.e. community-building. This approach is also applied towards the producers. The Manager for Produce Contact described it in the following way:

> When I began working here, I also started to call the producers [regarding the weekly orders], but it turned out to be very important that they speak to the [Manager for Producer Contacts]. He calls them and asks about their personal lives, they talk to him, and he listens. For instance, one of our producers has got a very bad lumbago, so we called him last week and said we're sorry about that.
>
> (Interview 2016b)

> We called him without a reason. So, it was not to ask "when can we expect you back?" But really, "How are you? We haven't heard from you for a long time".
>
> (Interview 2016a)

This shows how a premium is put on nurturing human relationships above professional ones, even if it is time-consuming and difficult to keep up in a growing organization. Features of producers, as well as plans to do the same for consumers, is a way to handle this. Kiskosár works on a membership basis, which comes with an expectation on customers and buyers not only to purchase products but also to be interested in the everyday life of producers and in linking up with them. The opposition to mere transactionism is shared by producers, as evidenced by a honey-producer interviewed for the study:

> I don't like to do it via shops but prefer to sell directly to customers. And here at Small Basket I thought I would find such a community [of buyers] who know what they want to buy, and I don't have to explain from scratch why honey is healthy, etc. This is a shopping community whose members have a conscious attitude to their eating habits.
>
> (Interview 2016c)

Shoppers, on their part, emphasized how trust in the quality of the product derived from the community-building:

> [Why do you come here?] The answer is very simple. Here, there is contact between the seller and the buyer. It is not just that you take down an item from the shelf that's impersonal, but here you know everyone.
>
> (Interview 2016d)

The Introduction to this volume emphasizes

> how the self-identifications of people in the borderlands may be an important asset and resource involved in attempts to deal with geopolitical changes to borders, and how processes of identity-formation might be understood as resources making borderlands resistant to change.
>
> (Andersen and Prokkola 2021, p. 000)

The Small Basket Community draws on several group identities to provide resources. First, it is part of a global movement promoting local food production and can thereby draw on the know-how developed regarding how to promote locally produced food as representing a more ethically responsible choice. Second, it can benefit from the economic nationalism underpinning policies supporting Hungarian production (Gilpin and Gilpin 1987, p. 31). But importantly,

> Small Basket Shopping Community differs from these two scales in that it deliberately markets a territory that spans a state border (the Hungarian-Slovak border) and therefore comprises citizens of two countries. This has often been on an ethno-linguistic basis – all Slovakian producers we met were Hungarian-speakers – and an added value of the Community can then

be seen as a way of strengthening Hungarian-Hungarian links in general, but also as a way of supporting the poorer Slovak side.

(Svensson et al. 2019, p. 54)

The navigation of these different scales takes skill and capacity to adapt to different contexts. Self-identification takes place at a local level where actors seek to distinguish themselves from national elites, in addition to the ethno-nationalist underpinning of a shared language. At the same time, it is driven by the motivation to be independent from the logic of industrial food production and beliefs that local resilience may be possible in sustainable ways. Advancing along these elements demonstrates resilience over time, understood as a process, yielding itself to bouncing-forward development.

Resilience can also be seen in its more common form, namely as the capacity to cope (adjust and mitigate) with sudden external shock. The closed borders of the COVID-19 pandemic impacted Kiskosár as well. In the spring of 2020, Slovakian producers had difficulties getting their produce across the borders, Slovakian shoppers could not pick up their weekly orders, and the mini-market format of the orders had to be organized to allow only those who pre-ordered into the premise and to open up for exact pick-up times to avoid queues. This worked contrary to the community-building element of these events and threatened the long-term survival of the food community. However, the organization could turn this situation to an advantage, demonstrating creative resilience going beyond either a bouncing-back or bouncing-forward narrative.

> The extraordinary situation of 2020 actually had some good effects for us. We always wanted to promote pre-ordering, but not everybody wanted to follow this. Now we could communicate very strongly that only people with pre-registered order could come to the collection points, which led to a big increase in our registered buyer number and the number of orders.
>
> (Interview 2020f)

In the fall of 2020, the cross-border dimension of the community had also stabilized somewhat, with the Slovakian producers managing to cross the border and some customers being able to make the trip as well. Yet it was clear that it would take long-time effort to fully regain what this crisis had added to the previous bureaucratic difficulties in the cross-border dimension of the enterprise.

Cross-border commuting: Securing livelihoods in growing agglomerations and neglected peripheries

Contrary to cross-border food communities, cross-border commuting has been a key interest of both policymakers and scholars, and the extent to which it takes place has been researched extensively. In the Hungarian context, cross-border commuting as a way of securing livelihoods both in growing cross-border agglomerations and in neglected peripheries became possible at some borders following the regime change in 1989/1990, and later through Hungary's EU accession in 2004 making its borders to Slovakia, Austria and Slovenia internal EU borders. Today the borders to Romania and Croatia are also internal EU borders, and Ukraine and Serbia are Hungary's only non-EU neighbors.

Consequently, Hungary is one of the EU countries with the highest cross-border out-commuting rates; with almost 2.5 percent of the working population commuting, the country scored a sixth place in 2018 (Meninno and Guntram 2020, p. 88). Austria remains significantly richer than Hungary, which largely explains why 83 percent of Hungarian cross-border workers were heading there (KSH 2015, p. 23). In turn, Slovakia and Romania host particularly large ethnic Hungarian communities, although these are by no means the only substantial groups of daily commuters. On the contrary, the large chunk of residential migrants in some of Hungary's rural borderlands are ethnic Slovaks and Romanians who recently moved there from larger cities located on the other side (Bratislava, Oradea, Košice, Satu Mare and Arad). Such urban agglomerations expanded as middle-class families (mostly not ethnic Hungarians) moved to rural areas on the Hungarian side, pumping new life into what earlier were mostly deprived and shrinking small towns. They developed a cross-border lifestyle that includes daily commuting back to Romania or Slovakia, where contacts (e.g. social, but also work-related) can be maintained on one side and newly created on another. There are also groups of commuters coming into Hungary (ethnic Hungarians and non-Hungarians alike) from Ukraine, Romania, Serbia and to a lesser extent Slovakia.

Intuitively, this process would not be described as resilience in this volume's and the present chapter's conceptualization of resilience as creating opportunities "when faced with adversity" and when a community is "disturbed" (Andersen and Prokkola 2021). This has to do with the largely positive narratives that surround how the regime change and EU accession led to the opening of borders. Opening borders is usually not framed as adversity or disturbance. However, by disentangling the various components involved, it may be interpreted as resilience, especially when focusing on commuters that do not move far. In a European context, East Central Europe remains economically peripheral, and massive out-migration to Western Europe and to regional growth hotspots has led to severe labor shortages and demographic decline. Those not aiming to emigrate far from their native regions have strong incentives to take advantage of all kinds of opportunities offered by the border location. The most obvious example is the daily or weekly commuting of labor to proximate places, often practiced by people with lower educational attainments than the average for Hungary's workforce (KSH 2015, p. 5). The close-distance migration implied in cross-border residential mobility can be connected to resilience in at least two ways. One, it can be seen as avoiding emigration to regions further away, thus maintaining local communities and relationships with relatives and friends. Secondly, while the opportunity for urban agglomerations to expand beyond the border does imply increased traffic between these new transborder suburbs and the centers of gravity, they also mean less congestion and a slower growth in housing prices than had these cities been limited to the confines of their host states. Rural Hungarian borderlands, which until recently were losing population, have received thousands of new residents, contributing to their ability to keep running local services like kindergartens and schools (Lovas Kiss 2018; Balizs and Bajmóczy 2018).

In these borderlands, open borders and transnational lifestyles have for many become the new normal, a normality that in turn generated a need of resilience in the face of temporary border restrictions and border closures, i.e., externally induced shocks to these territories in the form of policy responses to the increase in refugee numbers in 2015 and the COVID-19 pandemic. Already during the first wave of the pandemic, Hungary was among the first EU countries to close its borders in mid-March (Rettman 2020), leading to long waiting times and confusion, especially along the border to Austria (Határátkelő 2020). In early June, Hungary, Slovakia, Czechia and Austria mutually agreed to open their shared borders, with the Slovak Prime Minister declaring that "we can now return, at least to some extent, to normal life" (Portfolio 2020). Hungarian measures during the second wave were similarly coupled with anxieties and uncertainties among cross-border commuters. In late August, Hungary declared that foreign citizens could not enter the country from September 1, except for well-founded reasons (Határátkelő 2020). Yet local cross-border communities, to some extent backed up by their national decision-makers, have thus far found ways to avoid a collapse of the sorts of lives they have been conducting over the past years or even decades. By the day of the regulation entering into force, it was modified to exempt Czech, Polish and Slovak citizens who had booked accommodation in Hungary for the month of September. The European Commission warned Hungary that discrimination among EU citizens goes against Union regulations (Ibid.), but the Hungarian government nevertheless prolonged the exemption for another month (About Hungary 2020).

Indeed, during both waves of the pandemic Hungary exempted persons residing in a zone of 30 kilometers from its Schengen borders from not crossing freely (yet still passing border checks and showing documents), practices used before to define 30-kilometer border zones with special rights to cross. Nevertheless, in early September a chaotic situation emerged when Hungarian authorities closed the smaller crossing points (Határátkelő 2020). As one result, for a few days Slovakian pupils attending high school in Balassagyarmat in northern Hungary had to travel 70 kilometers instead of just a few, an issue whose gravity is reflected by the fact that thousands of Slovakian pupils and students are registered at Hungarian schools (Pregi 2018) and universities, including 2,000 at university colleges alone (Határátkelő 2020).

Reactions to the implemented Hungarian measures quickly arose from a wide range of actors on different scales. For instance, Slovakia's State Secretary for Education unofficially approached Hungarian authorities to reconsider the regulations (Finta 2020). Further, to dampen uncertainties media portals specifically targeting borderland communities intensively reported on the latest developments and experiences of trying to cross the border. In one such video report, affected persons are interviewed, emphasizing the difficulties and challenges of crossing, especially beyond the designated 30-kilometer zone (Paraméter 2020). As an interviewed old lady commented, "we have been locked in again . . . but might try to pass at night." It appears that such rule-resistance is a more general way borderlanders deal with the stress of recent border closures (cf. Lois et al. 2022), however, for local cross-border cooperation organizations, such as Euroregions and European Groupings of Territorial Cooperation (EGTCs), the changed circumstances constituted a challenge in terms of aligning with national policies while trying to minimize disruptions to citizens on both sides of the borders. According to representatives of four cross-border organizations at the borders with Croatia, Serbia and Slovakia (Interview 2020a, 2020b, 2020c), the bilateral dialogues regarding the general situation of cross-border commuters were generally conducted without the participation of local organizations. However, in some important instances, local organizations could advocate for the visibility and solution regarding specific problems, such as the operation of a ferry or farmers' mobility.

> We are in quite close cooperation with the foreign ministry and signaled [the problem] to them.
>
> (Interview 2020b, Cross-border cooperation
> organization representative)

> It was only for a couple of days that no one could cross. After that commuters who could show a stamped and signed paper proving their employment could cross. This mainly concerned Slovakian residents working in Hungary, since the workflow goes that way. [Was this due to your actions?] I don't think so, it [the decision] took place at the national level, but there was another issue [regarding a bridge] about which the EGTC contacted the Foreign Ministry, which was very helpful and within a week they had a solution.
>
> (Interview 2020c)

At the European level, the cooperation around solving the commuting issue at Hungary's borders was even promoted as best-practice in terms of cooperation (see for instance Committee of the Regions 2020), but not all borders were equally impacted. For instance, the situation was slightly less dramatic along the Austrian-Hungarian border, where guards were less keen on checking documents during those rainy September days. This is important, for as a dentist in Szombathely commented, "without the Austrian clients we might as well close" (Határátkelő 2020). The same is true in the other direction. As Hungarian employees at Caritas Burgenland commented: "Since the start of the coronavirus we have learned how to react flexibly to emerging situations. Thus, we now have a plan B for emergency situations, but so far we don't need to take any measures of precaution" (Ibid.). Later, Hungarian decision-makers had listened to such articulated concerns and have re-opened the suddenly closed borders during September (Szász 2020). This type of learning among actors showcases a process that also interlinks different scales, going beyond administratively bounded territories at the respective side of the border. It can be discussed whether the selective lobbying activity for mitigating the effects of border closures and to resist full closures indicates only bouncing back, rather than forward-bouncing resilience, but it definitely encapsulates the notion of reactions to adversity and disturbance.

Cross-border solidarity and humanitarianism: Resisting the rhetoric and practices of closed borders and societies

After having investigated resilience at sites of food community building and cross-border urbanization, we now turn to resilience against ideas around the benefits of "secure" and "strong" borders. While many practitioners working with cross-border cooperation in Europe over the past decades have assumed that physical manifestations of borders (crossings, controls, and walls) are fundamentally negative, the tacit support for "border management" in Europe has shown that people can also perceive border security as a positive change. In Hungary, such sentiments have been officially promoted under the Fidesz-led government, with especially the idea of borders as protection against migrants playing a dominant role. Indeed, in Europe the years of 2015 and 2020 have become shortcuts, synonyms or symbols of bordering, with 2015 representing bordering as a reaction to refugee influx, and 2020's closed borders as a reaction to a previously unknown disease, COVID-19.

In 2015, more than a million refugees arrived in Europe in a way that was perceived as uncontrolled and unorderly by many of its citizens. This led to stricter border controls and challenges to the cross-border cooperation structures at many European border areas (Engl and Wisthaler 2020; Prokkola 2019; Svensson 2020). The situation in Hungary was more dramatic than in most EU states because of a combination of being one of the major transit countries and having a government with a pronounced anti-immigrant and anti-refugee stance. It became the site of international attention with dramatic footage from Budapest railway stations where refugees were held up, and the construction of a fence at Hungary's southern borders to Serbia and Croatia. Still, it was possible for local activists to show solidarity and provide material support for transiting refugees while also building small local communities around the idea of resistance (Svensson et al. 2017), here outlined as the case of MigSzol Szeged.

Szeged is a city of 162,000 inhabitants at Hungary's border to Serbia, where an action group called MigSzol Szeged provided humanitarian assistance to an estimated 30,000 refugees in 2015. MigSzol Szeged never registered as a formal non-governmental organization, but it mobilized around 150 citizens in the work to provide food, clothing and other assistance to people desperate to continue northwards to countries such as Germany and Sweden. Szeged is considered somewhat of an opposition town in Hungary to the national-conservative government Fidesz and has had a mayor from the socialist opposition party. This helped contacts between MigSzol Szeged and local public administration, and the group received help in constructing a local humanitarian hotspot, demonstrated by the following quotes representing first the public administration, and then the civil society group.

> I have to say they [MigSzol Szeged] organized themselves extremely well; they had to organize the changing of their staff, distributing donations from Germany etc.; thus serious logistical abilities were manifest. After seeing the railway station, the activists approached us at the city council if we could help with for instance some wooden stands at the station, at which they could distribute food, some storage and toilets.
>
> (Interview 2016e)

> It was amazing that at one pm we got out from the Mayor's office, where they took note of what we wanted, adding that if they really hurry they could have it all within a week, but the next day at noon the wooden cottage was there, the electricity worked, etc. 23 hours after our meeting they put it all together, including necessary permissions.
>
> (Interview 2016f)

The good relations were ascribed both to the political color of the leadership and to the town spirit.

> I have a simple answer to this [why Szeged cooperated more with civilians]: we are better people. Hundreds of citizens helped people who they knew were walking 2,–3,000 kms and are exhausted. I think it's normal that the city helped, too. These people [civilians] assumed duties of the state. The Hungarian state could build hundreds of kilometers of fences, and within a few hours organize some buses to – btw illegally – transport people from one of its borders to another, but it is unable to print some info flyers and maps.
>
> (Interview 2016e)

The relation with others in the local population was not always as good, with several members of the group being exposed to verbal harassment and hate. Mig-Szol Szeged members generally emphasized the humanitarian side of their work and did not acknowledge its potential political nature.

The several-month-long voluntary work with refugees created strong bonds between the volunteers, and the core group continued its work after the situation had peaked, trying to assist refugees in the newly erected border camps in Hungary and Serbia as long as they could. A group of local citizens had been created, who were active both in Hungary and across the border, sharing and strengthening specific norms with regards to mobility, thereby creating common ground for resilience against national anti-migrant narratives. Thus, within a very short period, what started as a private initiative transformed into a large-scale humanitarian aid effort. In the lives of those they helped, this was just one or a few days of a long and arduous journey, but undoubtedly the personal impact of that moment was significant. The volunteers themselves were often profoundly affected by the events that unfolded. A sense of community was swiftly created, a we-feeling often against what was perceived as a hostile society. Nonetheless it is noteworthy that this hostility did not extend to the local authorities, with which relations were perceived as good and productive.

Organizations dedicated to cross-border cooperation, such as Euroregions and European Groupings of Territorial Cooperation mentioned earlier, were generally not involved in policy action in 2015, which they saw as national prerogatives, something that is not unique to actors at Hungarian borders (Svensson 2020). The situation in 2020 was different: instead of selective border measures to keep certain groups of people out, initially everyone was hindered to cross through blanked border closures and lockdowns. This threatened the very essence of local cross-border cooperation, yet it opened up for an upscaling of humanitarian initiatives. For instance, an organization at the Hungarian-Serbian-Romanian border quickly arranged a humanitarian action, providing protection gear (masks) and disinfectants to 37 Romanian municipalities that were all members of a cross-border cooperation body.

> Within one week, this decision was taken, and it was a quick movement to realize this. We sent the items with a commuter who could cross the border, and it was a very positive reception. They were very grateful. They had not gotten any help, even minimal, from anyone, so they could see that they got something positive from being in this organization.
>
> (Interview 2020a)

Andersen and Prokkola (2022) emphasize that there is a lack of research on how

> political borders and strengthened border securitization hinder the vernacular and regional resilience strategies in the face of environmental, economic and socio-cultural change. When border communities and mobile people need to cope with manmade material border infrastructures, renewal and resistance may emerge as a response to such border transitions.
>
> (p. 5)

This section demonstrated the existence, even if marginal, of resilience that goes beyond pragmatic coping, adaptation and flexibility, and instead capturing processes that create and sustain resilience of ideas, a process that is inherently politically charged in a context where borders themselves become politicized. Inspired by Rogers et al. (2020), we argue that resilience is understood differently depending on ideological orientation. Those adhering to the value of keeping the Hungarian nation together, partly by promoting mono-ethnic encounters and cooperation formats, are supportive of borders and border controls as a means to prevent "the Other" (foreigners, migrants) as a way to increase the resilience of "the nation." At the same time, they are confronted with the dilemma of keeping borders porous enough to allow for mono-ethnic cross-border cooperation, which is perceived as increasing resilience. Others countered this narrative and instead chose to support refugees as a civil society resistance.

Conclusion

This chapter investigated food community creation, commuting, and solidarity activities taking place at Hungarian border sites as three cases through which we could analyze resilience as a process. We first illustrated through the development of a local transborder food community how a movement that spans multiple scales shows resilience over time, while reactions to the COVID-19 crisis showed the need for more old-fashioned bouncing-back resilience. We then looked at cross-border commuting patterns in tandem with the creation of cross-border agglomerations, arguing that understanding this as resilience depends on time-perspective. This finding fed into the study of cross-border solidarity and humanitarian actions towards refugees as direct responses to hardening borders, and how this highlights a neglected political and value-related dimension of resilience.

The cases highlight that resilience may indeed be manifest as reactions to both multiple slow disruptions and sudden single disruptions. The analytical distinction between bouncing-back and bouncing-forward resilience could relatively easily be applied to the three cases, yet it did not prove very fruitful for providing enriched understanding of these processes. Our study is therefore an answer to calls for contextualized approaches that also consider the crisscrossing of various scales and bounded entities, in contrast to more rigid, "placeless" benchmarking measurements (Bristow 2010). The overall inquiry highlights how applying a process-based approach to resilience, also accounting for values and political resistance (Rogers et al. 2020), helps uncover developments in borderlands that may otherwise remain unearthed.

Note

1 Interviews from 2020 cited in this chapter were carried out by Sara Svensson during a fellowship awarded by the Robert Schuman Centre for Advanced Studies at European University Institute in Florence, Italy. In the case of co-author Péter Balogh, research for this paper has been implemented with support from the National Research, Development and Innovation Fund of Hungary (NKFI K 134903 Geopolitical processes and imaginaries in Central Europe: states, borders, integration and regional development). In addition, at the time of writing Péter Balogh was a beneficiary of the János Bolyai Research Scholarship of the Hungarian Academy of Sciences. Interviews from 2016 cited in this paper were carried out jointly by the authors for the Center for Policy Studies at Central European University (Budapest) within the project "SOLIDUS Solidarity in European Societies: Empowerment, social justice and citizenship," which received funding from the European Union's Horizon 2020 program, under grant agreement number 649489. See Svensson et al. (2017) for more details.

References

About Hungary, 2020. Travel restrictions amended for V4 countries. *About Hungary*, 6 October. Available from: http://abouthungary.hu/news-in-brief/travel-restrictions-amended-for-v4-countries/

Andersen, D.J. and Prokkola, E-K., 2021. Introduction: embedding borderlands resilience. *In*: D.J. Andersen and E-K. Prokkola, eds. *Borderlands resilience: transitions, adaptation, and resistance at borders.* London: Routledge, 1–18.

Balizs, D. and Bajmóczy, P., 2018. Szuburbanizáció a határon át: társadalmi, etnikai és arculati változások Rajkán [Rajka: the "Hungarian suburb" of Bratislava]. *Tér és Társadalom,* 32 (3), 54–75.

Balogh, P. and Pete, M., 2018. Bridging the gap: cross-border integration in the Slovak–Hungarian borderland around Štúrovo–Esztergom. *Journal of Borderlands Studies,* 33 (4), 605–622.

Benke, M., Czimre, K., Forray, K.R., Kozma, T., Márton, S., and Teperics, K., 2018. Learning regions for resilience in Hungary: challenges and opportunities. *In*: T. Baycan and H. Pinto, eds. *Resilience, crisis and innovation dynamics.* Cheltenham: Edward Elgar, 68–89.

Bristow, G., 2010. Resilient regions: re-'place'ing regional competitiveness. *Cambridge Journal of Regions, Economy and Society,* 3 (1), 153–167.

Committee of the Regions, 2020. We stand together. cities & regions responding to the Covid-19 emergency. Available from: https://cor.europa.eu/fr/engage/Pages/covid19-stories.aspx

Davies, S., 2011. Regional resilience in the 2008–2010 downturn: comparative evidence from European countries. *Cambridge Journal of Regions, Economy and Society,* 4 (3), 369–382.

Diaz-Sarachaga, J.M. and Jato-Espino, D., 2019. Do sustainable community rating systems address resilience? *Cities,* 93, 62–71.

Engl, A. and Wisthaler, V., 2020. Stress test for the policy-making capability of cross-border spaces? Refugees and asylum seekers in the Euroregion Tyrol-South Tyrol-Trentino. *Journal of Borderlands Studies,* 35 (3), 467–485.

Fazekas, N., Fábián, A., and Nagy, A., 2017. Analysis of cross-border regional homogeneity and its effects on regional resilience and competitiveness with the Western Transdanubian region (HUN) and Burgenland (AUT) as examples. *Acta Universitatis Sapientiae, Economics and Business,* 5 (1), 5–28.

Finta, M., 2020. Enyhítést kérnek a határon. *Új Szó,* 1 September. Available from: https://ujszo.com/kozelet/enyhitest-kernek-a-hataron

Gilpin, R. and Gilpin, J.M., 1987. *The political economy of international relations.* Princeton, NJ: Princeton University Press.

Határátkelő, 2020. Felemás határzár, kétségbeesett ingázók. *Határátkelő,* 3 September. Available from: http://hataratkelo.com/felemas-hatarzar-ketsegbeesett-ingazok/

Interview, 2016a. Manager for producer contacts and products. Kiskosár bevásárló közösség (Small Basket Shopping Community). Esztergom.

Interview, 2016b, 2020d. Manager for daily operations and the volunteer organization, Kiskosár bevásárló közösség (Small Basket Shopping Community). Esztergom, 10 March 2016. Email follow-up 27 November 2020.

Interview, 2016c. Producer at Kiskosár bevásárló közösség (Small Basket Shopping Community). Esztergom, 10 March 2016.

Interview, 2016d. Long-term member of the Kiskosár bevásárló közösség (Small Basket Shopping Community). Esztergom, 10 March 2016.

Interview, 2016e. Sándor Nagy, Deputy mayor of Szeged, Szeged, 7 June 2016.

Interview, 2016f. Márk Z. Kékesi, MigSzol Szeged coordinator (communications). Szeged, 6 June 2016.

Interview, 2020a. Cross-border cooperation organization representative. *Online Interview*, 22 June.

Interview, 2020b. Cross-border cooperation organization representative. *Online Interview*, 24 June.

Interview, 2020c. Cross-border cooperation organization representative. *Online Interview*, 26 June.

Jones, R., 2016. *Violent borders*. London: Verso.

KSH [Központi Statisztikai Hivatal], 2015. Ingázás a határ mentén. *Központi Statisztikai Hivatal*. Available from: www.ksh.hu/apps/shop.kiadvany?p_kiadvany_id=80484&p_temakor_kod=KSH&p_lang=hu

Lendvai, E. and Ördög, Á., 2018. Analysis of the Hungarian agricultural marketing – by the supply of local products' consumers. *Analecta Technica Szegedinensia*, 12 (2), 37–44.

Lendvay, M., 2016. Resilience in post-socialist context: the case of a watermelon producing community in Hungary. *Hungarian Geographical Bulletin*, 65 (3), 255–269.

Lois, M., Cairo, H., and García de las Heras, M., 2022. Politics of resilience . . . politics of borders? In-mobility, insecurity and Schengen "exceptional circumstances" in the time of COVID-19 at the Spanish-Portuguese border. *In*: D.J. Andersen and E-K. Prokkola, eds. *Borderlands resilience: transitions, adaptation, and resistance at borders*. London: Routledge, 54–70.

Lovas Kiss, A., 2018. A határon átívelő lakóhelyi mobilitás az ártándi határátkelőhöz kapcsolódó forgalmi folyosó mentén, Románia európai uniós csatlakozását követően. *Erdélyi Társadalom*, 16 (1), 151–168.

Meninno, R. and Guntram, W., 2020. As coronavirus spreads, can the EU afford to close its borders? *In*: R. Baldwin and B. Weder di Mauro, eds. *Economics in the time of covid-19*. London: CEPR Press, 87–91.

Paraméter, 2020. Komáromban mindenféle ellenőrzés nélkül át lehet jutni a határon (Videó). *Paraméter*, 4 September. Available from: https://parameter.sk/komaromban-mindenfele-ellenorzes-nelkul-lehet-jutni-hataron-video

Pirisi, G., 2017. A reziliencia szerepe a kisvárosok fejlődésében – egy komlói esettanulmány kapcsán. *Településfejlesztési Tanulmányok*, 6 (2), 75–88.

Portfolio, 2020. Képeken a szlovák határ teljes megnyitása. *Portfolio*, 5 June. Available from: www.portfolio.hu/gazdasag/20200605/kepeken-a-szlovak-hatar-teljes-megnyitasa-435600

Pregi, L., 2018. A magyarországi közoktatásban részt vevő szlovák állampolgárok területi megoszlása [Spatial pattern of Slovak nationals participating in public education in Hungary]. *Területi Statisztika*, 58 (2), 151–176.

Prokkola, E-K., 2019. Border-regional resilience in EU internal and external border areas in Finland. *European Planning Studies*, 27 (9), 1587–1606.

Rettman, A., 2020. Nine EU states close borders due to virus. *EUobserver*, 16 March. Available from: https://euobserver.com/coronavirus/147742

Rogers, P., Bohland, J.J., and Lawrence, J., 2020. Resilience and values: global perspectives on the values and worldviews underpinning the resilience concept. *Political Geography*, 83.

Sensier, M., Bristow, G., and Healy, A., 2016. Measuring regional economic resilience across Europe: operationalizing a complex concept. *Spatial Economic Analysis*, 11 (2), 128–151.

Shaw, K., 2012. The rise of the resilient local authority? *Local Government Studies*, 38 (3), 281–300.

Sohn, C., 2014. The border as a resource in the global urban space: a contribution to the cross-border metropolis hypothesis. *International Journal of Urban and Regional Research*, 38 (5), 1697–1711.

Svensson, S., 2020. Resistance or acceptance? The voice of local cross-border organizations in times of re-bordering. *Journal of Borderlands Studies*. Epub ahead of print 3 July.

Svensson, S., Balogh, P., and Cartwright, A., 2019. Unexpected counter-movements to nationalism: the hidden potential of local food communities. *Eastern European Countryside*, 25 (1), 37–61.

Svensson, S., Cartwright, A., and Balogh, P., 2017. Solidarity at the border: the organization of spontaneous support for transiting refugees in two Hungarian towns in the summer of 2015. Center for Policy Studies Working Paper Series, 1/2017. Budapest: Central European University.

Svensson, S. and Nordlund, C., 2015. The building blocks of a Euroregion: novel metrics to measure cross-border integration. *Journal of European Integration*, 37 (3), 371–389.

Szász, P., 2020. Sorban nyitotta meg a korábban lezárt határátkelőket a kormány. *Napi.hu*, 22 September. Available from: www.napi.hu/magyar_gazdasag/sorban_nyitotta_meg_a_korabban_lezart_hataratkeloket_a_kormany.714129.html

Wandji, G., 2019. Rethinking the time and space of resilience beyond the West: an example of the post-colonial border. *Resilience: International Policies, Practices and Discourses*, 7 (3), 288–303.

6 Mobility turbulences and second-home resilience across the Finnish-Russian border

Olga Hannonen

Introduction

Borders and border institutions provide an illustrative and dynamic context to examine resilience. The year 2020 has showed that borders can rapidly transform from open to closed, creating abrupt and exceptional circumstances for border-landers and border crossers. In Finland, for example, the guidelines on border-crossing have changed every month since March 2020. Border and mobility regulations affect populations at large, demonstrating that borderlanders are not only those who reside in proximity to the border. A borderlander can be anyone who has connections to places across borders in the form of property owner-ship or family or other personal or professional connections. Thus, as empha-sized by Andersen and Prokkola (2022), border studies have great potential in contributing to understanding of social resilience. "Social resilience" here refers to the ability of human communities to withstand various stresses (Cheer and Lew 2017). To further develop this perspective, this chapter brings a detailed overview on social border resilience in the context of changing cross-border mobilities and mobility regulations. It focuses on a distinct group of travelers and semi-permanent residents in the Finnish border region, Russian second-home owners. Though rather specific, the case is illustrative of how changes to border and mobility regulations have stronger impacts on second-home owners com-pared to average visitors to the border area. Those persons who have a permanent attachment to a place and perform regular visits to the area face a wider spectrum of regulations and policies. While average visitors can simply decide between a visit or no visit, property owners need to find creative adaptive ways to navigate within the changed circumstances.

To examine mobility across the border and mobility regulations, the chapter employs the concept of *mobility turbulence* (Cresswell and Martin 2012). Tur-bulence is defined as a "movement out of permanent order" (Cresswell and Mar-tin 2012, p. 520). It conceptually frames changing mobility conditions across the Finnish-Russian border and the resilience practices as a social response to these processes. The notions of turbulence and resilience often appear together in studies on corporate and market resilience in times of economic instability. Surprisingly in mobilities research and social science at large, these two concepts

DOI: 10.4324/9781003131328-8

have barely intersected. In response, this chapter conceptually bridges mobility, border and resilience studies. In tourism research, the issues of crisis, disaster and security have not been strongly linked to the resilience literature (Hall 2018). Thus, the chapter contributes to tourism studies and studies on social border resilience.

The chapter offers a grassroots perspective on Russian cross-border second-home mobilities to Finland under changing mobility regulations after the year 2013. The main research question is: "What resilient solutions do Russian second-home owners create under the changed mobility circumstances?" While global geopolitical and economic events are often considered as the most challenging changes, others such as legislative changes and refinement of regulations might lead to greater adaptation and resilience in the lives of mobile people and/or borderlanders.

Since 2013, Russian second-home mobility to Finland has undergone a series of turbulences that transformed physical relocations across the border, cross-border economic activity and property purchase practices. The discussion in this chapter focuses on mobility sanctions enacted after the Ukrainian conflict in 2014, modifications to Schengen visa regulations, changes to banking policies in Finland and Russia, and security discourses and legislative change in relation to Russian property purchases in Finland. To understand the resilient solutions of Russian second-home owners under the changed circumstances, the chapter employs interviews with Russian second-home owners in the Finnish border regions of South Savo and South Karelia. These two regions are the most popular among Russian property purchases (Hannonen 2016). The information on foreign purchases in the regions was provided by the National Land Survey of Finland. Altogether, 26 interviews with Russian owners were conducted in summer 2017. All interviews have been conducted by the author in Russian. The author was responsible for transcription, interpretation and translation of the interviews. Personal information about the respondents has been removed to preserve anonymity. The purpose of the interviews was to understand Russian second-home owners' mobility and everyday life in Finland after the changed socio-political and economic circumstances after the year 2013. The following section presents a conceptual discussion on social border resilience to position the case within the resilience framework.

Resilience, border and turbulence

The concept of resilience has several approaches that vary from a response to extreme change, gradual adaptation to changing conditions and means of resistance, to changes in values as a survival strategy (Amore et al. 2018; Brown 2016; Humbert and Joseph 2019). Cheer and Lew (2017) define 13 different types of resilience. Among those, *social resilience, fast change* and *slow change* are useful terms to define and categorize the resilience of Russian second-home owners. *Social resilience* refers to the ability of human communities to withstand various stresses. *Fast change* is defined as "change linked to sudden and often unexpected events that requires an immediate system response," while *slow change* is a gradual adaptation to long-term global changes (Cheer and Lew 2017, p. 9).

Social resilience is a conceptually flexible approach that can be used to define the adaptive strategies, changes and resistance of social groups and regions. Social dimensions are also present, for example, in regional resilience, which includes a sense of security among populations and institutional trust among local populations (Prokkola 2019, p. 1591). Sense of security and trust are regarded as important components in the attractiveness of regions to international tourists and second-home owners (Hannonen 2016; Lipkina 2013). The community perspective and regional and business resilience have been dominating perspectives in resilience research in tourism studies (see Cheer and Lew 2017; Saarinen and Gill 2018). At the community level, resilience is understood as an ability to adapt to adverse events (Hall 2018). Research examples include tourism-induced transformations and tourism development impacts, issues such as overtourism, overdevelopment, policy and planning, and environmental change among others (Amore et al. 2018; Hall 2018; Saarinen and Gill 2018). In contrast to these dominant perspectives, this chapter looks at tourists and their adaptive strategies under changing socio-political circumstances and border regulations. Therefore, the major focus in the analysis is on the human dimension in social resilience. Human elements in resilience are considered to integrate "a new set of ideas around adaptation and adaptive capacity, learning and innovation" (Brown 2016, p. 79). Thus, this can lead to personalized and creative trajectories of coping with change.

There is an ongoing interdisciplinary debate whether the extent to which resilience should be understood as the return to a previous equilibrium or reference state (e.g. Hall 2018) or as a transformation. In the case of border regions and borders at large, it is often extremely hard or impossible to trace or define the "optimal condition" to return to. When the example of the Finnish-Russian border is taken, which has been moving from a completely closed towards a more open and interactive border, it is not possible to define a reference state. That said, recent studies and the pandemic-driven border closures have shown that a return to a state of closure and limited interaction is not a desirable scenario (Gurova and Ratilainen 2016; Pitkänen et al. 2020; Prokkola 2019).

With respect to social groups, resilience can be "integrated into diverse ways of thinking about the human condition and the domestication of geographical living spaces" (Wandji 2019, p. 288). It has positive and negative connotations, encompassing adaptation as well as problems in that process. Wandji (2019, p. 288) states that since "varying forms of resilience exist," the application of a resilience lens is a situational and contextual project. Indeed, when we talk about borders, it is important to point out that borders are not just fixed physical structures but multi-layered constructs that encompass political, economic and cultural barriers. Furthermore, borders create a specific context of constant change and flow that triggers resilience. Prokkola (2019) has introduced the concept of *border-regional resilience*. She argues that "geopolitical situation and formal and informal border institutions partly determine the modes of adaptation and coping mechanism" (Prokkola 2019, p. 1588). Thus, border-regional resilience is not exclusively triggered by changes in border and/or mobility regulations. Other intervening factors, such as economic decline, political tensions and sanctions, and recently pandemic-driven nonpharmaceutical interventions in the form of health checks and quarantines, affect mobility across the border (Hannonen and Prokkola, in press).

In a study on borderlanders' responses to disruptive practices across the Gabon-Cameroonian border, Wandji (2019) argues that by examining people's everyday life and practices as a response to external disruption, we can not only construct a descriptive example of social resilience in a particular context but also co-construct the notion of resilience through empirical realities. In a similar vein, Andersen and Prokkola (2022) state that it is important to pay attention to how borderlanders adapt to stress situations, "how they interpret, articulate and make sense of different shocks, and how this influences their responses." This chapter follows these perspectives on co-creating resilience as a collective empirical construct by looking into the adaptive practices of Russian second-home owners.

Empirical studies on border resilience are conducted under circumstances of specific change, which is often regarded as "mobility shock" or "external disruption" (Prokkola 2019; Wandji 2019). Thus, taking into account previous elaborations on the subject, this chapter proposes additional conceptual facets of resilience based on the example of Russian second-home mobility in Finland. The chapter employs the concept of *mobility turbulence* (Cresswell and Martin 2012) that is used in mobilities research to define "movement out of permanent order" (Ibid., p. 520).

Turbulence is understood both as turbulence *within* mobility as well as turbulence *for* mobility (Cresswell and Martin 2012). Unexpected political events, catastrophes and natural disasters create turbulence *for* and *within* mobilities. In a border context, the opening or closure of borders is a turbulence that modifies, produces or restricts cross-border moves. However, the understanding of turbulence should go beyond its notion as an exceptional state of disorder; rather, it can be approached as a starting point for understanding movement and mobilities (Cresswell and Martin 2012). In such a manner, turbulence and resilience can symbolize positive and creative processes.

Cresswell and Martin (2012, p. 520) argue that "correct mobilities" are produced through the designation of routes. Thus, border infrastructures and international regulations orchestrate mobilities and flows in an accepted, smooth or non-turbulent manner. However, the empirical case of Russian second-home owners in Finland demonstrates that these regulating and structuring measures can in turn lead to turbulence *within* mobilities and abruptly change mobility patterns and practices – all of which lead to resilience in the form of adaptation and adjustment to the changed circumstances.

Not all changes and shocks can be negotiated through resilience. The nature of the change predefines the leeway for social resilience. Thus, I propose to examine the capacity for adaption to border and cross-border mobility changes through

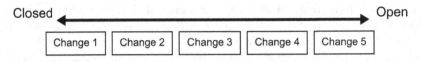

Figure 6.1 An open-closed resilience continuum.

an open-closed continuum, in which the open end gives space for resilient solutions, while the closed end does not give room for negotiation of changes in behavior or practices (Figure 6.1).

Thus, rather than searching for an equilibrium or reference state, I suggest that defining the capacity and degree of adaptation is a useful analytical tool in studying social border resilience. The resilience continuum is utilized in this chapter to define the degree of adaptation of Russian second-home owners in Finland to each mobility turbulence. To better understand how social border resilience is constructed during mobility turbulences in a border context, the next section analyzes the empirical case of Russian second-home owners' resilience in Finland.

Mobility turbulences across the Finnish-Russian border

The Finnish-Russian border is an illustrative case of changing regulations, mobility patterns and regional adjustments to increasing and decreasing cross-border flows (Hannonen 2016; Prokkola 2019). Most of the 1340-kilometer-long Finnish-Russian border runs through a sparsely populated or uninhabited northern wilderness. After 1944, the border with the Soviet Union was strictly controlled and functioned as part of "the East-West dividing line in Europe" (Eskelinen and Jukarainen 2000, p. 255), which has resulted in an economic peripheralization of the Finnish borderlands, especially those in the southernmost part of the border. These regions have weakly adapted to the changed circumstances, resulting in out-migration due to lacking economic possibilities. The Finnish-Russian border region has been defined as geographically and geopolitically sensitive (Hannonen 2016; Paasi 1995, 1999; Prokkola 2019). At present, the Finnish-Russian border is an external border of the European Union (EU) and, in many ways, has remained a hard and separating border.

The opening of the border for tourist visits has happened only after the collapse of the Soviet Union in 1991. Since then, Russian visits to Finland have been a growing trend: from 350,000 visits in 1991 to 3.9 million in 2019 (Atorus 2020). Finnish tourist visits to Russia comprised 896,000 in 2019 (Federal Agency for Tourism 2020). The southernmost stretch of the border (about 135 km long) between the Finnish regions South Karelia and Kymenlaakso and the Russian Leningrad region and the Federal City of St. Petersburg features the major Finnish-Russian interaction (trade, transit traffic, tourism etc.) (Smętkowski et al. 2016). Around 80 percent of all Russian visitors to Finland cross there. The inflow of Russian tourists has significantly transformed previously peripheral and declining border areas in Finland. For instance, "the influence of Eastern tourists has turned Lappeenranta from 'a slightly remote border town' into 'a giant shopping mall'" (Gurova and Ratilainen 2016, p. 58). Daily shopping tourism from Russia to the southeast border area of Finland became "the cornerstone of regional economic and tourism development" (Prokkola 2019, p. 1594). The exponential growth of Russian visitors from 1991 until 2014 has rapidly changed the landscape of local cities and put regional resilience to the test. To meet the demand, new shopping malls, spa centers, hotels and outlets were constructed in major population centers in the bordering areas. Such rapid changes have been welcomed by regional authorities. As a result, intensive cross-border interaction has not only contributed to economic development of border areas in the southeast of Finland, but also improved local attitudes towards Russian visitors (Gurova and Ratilainen 2016; Paasi 1999; Prokkola 2019).

In addition to tourist visits to Finland, second-home ownership has become a popular trend among Russians since 2000. Russians have rapidly outnumbered other foreigners on the Finnish property market. Currently, Russians comprise the biggest group of foreign second-home owners in Finland. According to the National Land Survey of Finland, Russians purchased 4,424 properties in Finland between 2000 and 2015, which is nearly twice the amount of all other foreign purchases for the same time period (N=2441). Russian property purchases account for 64.4 percent of all foreign property acquisitions in Finland between 2000 and 2015 (Hannonen 2016).

A border-imposed resilience is an omnipresent feature of Russian second-home ownership in Finland. Russian owners in Finland have demonstrated various strategies for adaptation that are driven by the socio-cultural and language differences. The lack of a common language, for example, has necessitated new manner of practical solutions in regard to the construction, purchase and maintenance of second homes and environments. These collectively have led to new service solutions in the form of a new real estate business sector (run by Russian migrants) that provides consultancy services and works exclusively with Russian clients (Hannonen et al. 2015). Another solution driven by the language barrier and lack of experience in the area is the creation of unique personal networks for service supply, consisting of a second-home seller, housebuilder or neighbors (Lipkina and Hall 2013).

The high level of cross-border activity initiated discussions on the creation of a visa-free regime that reached the top negotiation level of state governments in 2012. In 2014, the discussions were suspended, and borders and border control returned to the political agenda with the outbreak of the Ukrainian conflict and the introduction of sanctions (Hannonen 2016). The collapse of the ruble has resulted in a sharp decline of Russian visits, posing mobility shocks to tourist-dependent Finnish border regions in the southeast (Prokkola 2019). The global financial crisis in 2008 and economic decline in Russia due to drastically reduced oil prices and the outbreak of the Ukrainian conflict in 2014 have likewise resulted in a decline of Russian foreign property purchases. The absence of Russian tourists in the border regions has become a "regional misery" that was exacerbated by the halted regional-scale cross-border cooperation and export of certain goods (Prokkola 2019, p. 1597; Gurova and Ratilainen 2016). This resulted in a reconsideration of regional development policies and the development of regional resilience to better cope with the changing volume of tourist flows (Prokkola 2019). Simultaneously, Russian second-home mobility to Finland has undergone a series of turbulences that have led to various degrees of adaptation and resilience strategies.

Visa regime

The Finnish-Russian border marks the visa regime between the two states. Russian property owners are entitled to a multiple-entry, two-year Schengen tourist visa to Finland with a maximum stay of 180 days per year without the right to social services and health care. The calculation of the length of stay in the Schengen area underwent modifications in 2013. The permission to spend 180 days a year or "three months during six months" rule has been reworded to "90 days in any 180-day period" (Regulation (EC) No 610/2013). This minor change has significantly modified mobility patterns to Finland. Previously, one could freely distribute the permitted three-month stay during the six-month period in the Schengen area, including the possibility of an uninterrupted three-month stay. After the three-month stay and/or the six-month period expired, one could *open* another six-month period with the next entry and stay if desired three more months (not exceeding six months per year in total). After the change, every time prior to entry in the Schengen area a third-country national should count 180 days back from the planned day of entry and calculate the total number of days that have been spent in the Schengen area since the visa starting date. In other words, if one uninterruptedly spends 90 days in the Schengen area, (s)he must wait another 90 days (e.g. until the 180-day period is fulfilled) and only after that may enter the Schengen area again (cf. Virtasalo and Snieders 2013). One of the property owners explains the situation:

> The problem is that previously one could spend, for example 90 days from 1 January until 30 June, and then another 90 days randomly until December 31. Now, if one gets a visa on 1 January, and spends 90 days let us say from 1 April until 30 June, according to the new rule of counting back, one cannot enter for a single day until 31 September! Not in July, not in August, not in September! Not for a single day! Awfully inconvenient!
>
> (Female, Moscow, Russia, summer 2017)

While these changes provide a clear and unified method of calculation for the authorities, they have created much confusion and the need for additional planning and modifications of travel plans. This change significantly affected longer stays in the Schengen area, which are highly desired by second-home owners. The introduced change regarding the calculation of the lengths of stay greatly affect the mobility patterns of Russian second-home owners who, in contrast to other

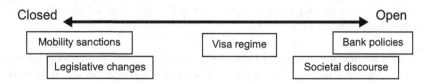

Figure 6.2 Degree of adaptation to mobility changes placed along the open-closed resilience continuum.

types of tourists, visit more frequently both for recreation and maintenance purposes. As a result, on top of the general limitation of the length of stay, Russian visitors need to distribute the allowed number of days according to the current formula. This change does not allow leeway or the possibility to find a creative way to adapt to the changed circumstances. Rather than resilience, it requires a *fast change* in behavior. While Russian owners are hopeful that the regulations will change to allow more flexible stays in Finland, their current visits are subject to accurate calculations and planning of entries, exits and duration of stay in any Schengen country. This change, whose goal was unified regulation and non-turbulent mobility, has resulted in turbulence *within* mobilities, disrupting and reorchestrating movements of Russian second-home owners. The formal acceptance of regulations does not develop resilience in the form of inner adaptation. While mobility has not been restricted in principle, the visa regulations in fact represent an imposed change. This means that while Russian visitors can exercise their mobility right under the visa regime, the format of the mobility requires practical modifications. Thus, modifications of the visa regime can be placed the middle of the resilience continuum (Figure 6.2), meaning that Russian owners can and must modify their mobility patterns under the changed circumstances, despite the limited scope of the change.

Mobility sanctions

After the outbreak of the Ukrainian conflict in 2014, the Ministry of Foreign Affairs of the Russian Federation issued a travel advisory to avoid travelling to countries, including Finland, which have an extradition agreement with the United States (Notice 2015). Shortly after that, about four million employees in various professions were prohibited from travelling abroad (cf. Hannonen 2016 for more details). Some of these restrictions were lifted, and currently the situation varies across different agencies. Simultaneously, Finland has joined the EU sanctions that restrict the entry of certain Russian citizens. The Ukrainian crisis has initiated legislative restrictions on foreign travel and property ownership for certain professional groups in Russia such as police workers (Draft Law N920563–6 2015), civil servants (Draft Law N624870–6 2014), and personnel of the Federal Guard Service (Order N168 2015; Order N154 2018). There is no precise information on whether these travel restrictions have impacted

Russian second-home owners in Finland. However, evidence shows that the families of some Russian second-home owners in Finland have been affected by these restrictions:

> It is all heading . . . toward building a barrier, as some professional groups cannot leave Russia. My daughter-in-law, she is a police worker, she is prohibited from going abroad, she simply cannot go to a second home [in Finland] with our family.
>
> (Male, St. Petersburg, Russia, summer 2017)

Similar to the changes in the visa regime, legislative changes cannot be circumvented. These turbulences *within* mobility necessitate *fast* change. Despite the second-home owners' perceptions and actions, these changes do not allow leeway for resilience, as cross-border mobility cannot be exercised by certain groups. Thus, they are located on the closed end of the resilience continuum (Figure 6.2). Mobility sanctions demonstrate another mobility shock in a series of turbulences *for* Russian second-home mobility in Finland.

Bank policies

Russian second-home owners have traditionally had access to a wide array of Finnish bank services, ranging from invoice payments in cash, bank accounts with cashless payments and property loans. In autumn 2016, news about the closure of Russian bank accounts in Finland had appeared in the Finnish media. While the banks did not provide reasons for their decisions, opinions in the press indicate that keeping an account merely for paying bills is not very profitable for banks, irrespective of the account holder's nationality (Salovaara 2016). However, the measures unevenly affected Russian owners, as according to the respondents, some of them have lost bank accounts; others have kept them under the same conditions.

The situation would not have been so acute if Finnish banks would not have simultaneously shifted to cashless payments. This means that banks no longer accept cash to pay bills, raising an issue of invoice payment for Russian owners. The remaining options are either to make expensive bank transfers from Russia to Finland or pay in cash at a Finnish chain of kiosks (R-kioski). The latter option, however, has been problematic, as according to one respondent it requires a Finnish ID number for utility payments. Both the termination of bank accounts and the shift to cashless payments have created a new type of invoice payment service among some real estate agents in Finland, who happily do this for an additional service fee. Another temporary solution is to turn to a neighbor or other acquaintance in Finland: "Our neighbors, he is a Finn, his wife is Russian. We ask them to pay online. This is the only solution" (Male, St. Petersburg, Russia, summer 2017). Turning to a neighbor or to an acquaintance for help in organizing maintenance services has been a common practice among Russian owners which previously related to the language barrier and lack of experience in the area

(Lipkina and Hall 2013). In case of bill payments, this is an unorthodox solution under the imposed circumstances.

Russian holders of Finnish bank accounts also do not have access to online banking in Finland: "We are never sure how much money we have in our account, which bills and what amounts have been paid automatically – we can roughly guess" (Female, St. Petersburg, Russia, summer 2017). The closure of the Finnish-Russian border in 2020 has made bank transfers even more problematic, raising the challenge of cross-border distance bank transfers for every owner.

Currently banks do not offer accounts to Russian property owners. While the author could not obtain any information regarding this issue from the Finnish banks, based on the information from the respondents and Finnish real estate agents, some banks require a Finnish co-signatory as a precondition to open a bank account. The decisions are, however, case-specific and inconsistent. None of the respondents in this study who had acquired property in Finland after 2014 and had met the co-signatory requirement could open a bank account:

> Every bill becomes a problem. We have to look for solutions. But it is stupid when you come with money and cannot pay a bill. We go to our main engineer, he luckily speaks Russian, and he takes care of our bills, but it is stupid when you have a bank [that could do it for you].
>
> (Couple, St. Petersburg, Russia, summer 2017)

The changed Finnish bank policies call for new solutions and adaptive measures, leading to the formation of mediator-practices. In addition to changes in Finland, in Russia owners have faced new regulations on foreign bank accounts and foreign property. Since 2015, Russian citizens have had to annually declare their foreign bank accounts, account activity and foreign property (Federal Law FZ-140 2015). The law obliges Russians to conduct all monetary activities through Russian bank accounts, defining money transfers to a foreign account as illegal activity, for example, the profit gained from property rent or money transfer from a property sale abroad. Not all Russian respondents were aware of the new legislation in Russia. Those who were aware are nevertheless reluctant to declare their assets:

> R: In Russia? Why declare it? The less they know the better.
> Me: So, you did not react to this?
> R: No, there is no need.
>
> (Male, St. Petersburg, Russia, summer 2017)

The mobility of assets across the Finnish-Russian border is an example of a turbulence *within* mobilities that leads to both resilience and denial practices (refusal to follow the rules that do not align with one's perceived life conditions can also be referred to as *act of refusal* (see Ferdoush 2022, p. 108). The Finnish side pushed for closure of bank accounts, creating resilient semi-temporary payment solutions that are placed on the open end of the resilience continuum (Figure 6.2). The

Russian side limited the mobility of assets of Russian citizens: the use of foreign bank accounts and money transfer. Thus, the law on declaration of foreign assets has been met with a *slow* response, or no response, and with denial. Denial is a form of resistance to change that has been examined as a response to border transitions (see Andersen and Prokkola 2022). This expands the presence of the border in the everyday, extending the definition of borderlanders. Similarly to entry restrictions, the legislative change on declaration of bank accounts in Russia does not provide much space for social border resilience; thus it is located on the closed end of the continuum (Figure 6.2).

Societal discourse and new regulations on property purchases

Russian land ownership in Finland has been contested at different levels, ranging from negative publicity and uneasy attitudes among locals, to legislative initiatives to restrict Russian property purchases in Finland (Hannonen 2016, 2020; Pitkänen 2011). Soon after they entered the Finnish property market, Russian owners became the subject of lively coverage in the national press, and Russian property purchases were colored by increasingly nationalistic rhetoric (Pitkänen 2011).

Russian property purchases have long been connected to national security concerns and discussions. However, with the outbreak of the Ukrainian conflict and the subsequent annexation of Crimea in 2014, such security discussions have become more heightened (Hannonen 2020). Russian property purchases have been viewed as potentially posing a threat to Finnish national security due to their possible location next to areas of national importance and the possibility of the Russian state seeking to protect its citizens across the border (PTK 16/2014; Vihavainen and Laitinen 2018). While strategic areas are not explicitly defined in the law, the areas in which the state has the primary right to oversee property purchases include: areas of importance for Finnish Defense Forces and the Border Guard, radar stations, harbors, airports, signal intelligence operational platforms and those areas that serve the Defense Forces and Border Guard's sea and air mobility, and other Defense Forces and Border Guard properties (HE 253/2018vp).

Security concerns came to the forefront in 2018 after a police raid on the Airiston Helmi Ltd. corporation premises located in the Turku archipelago in Western Finland. Several properties in the archipelago are located in close proximity to Finnish deep-sea routes and two army exclusion areas that are defined as areas of strategic importance. The reason for the raid was suspicion of white-collar crime, tax violation and money laundering. Though Airiston Helmi corporate property ownership should not be equated with Russian second-home purchases, in the public and parliamentary discourses the two phenomena are often conflated (cf. SKT 144/2018). The case has resulted in numerous discussions in the parliament on the need for new legislation.

As a result, a new law concerning national security considerations in property purchases and area use was enacted in January 2020. Every property purchased by a non-EU or non-EEA national must be inspected on location if it is in the vicinity of strategically important objects or areas. This change subjects each purchase to a license issued by the Ministry of Defense (HE 253/2018vp). The new law concerns future purchases; thus at the moment it is unclear whether and how properties in current Russian possession will be inspected, and what would happen to those properties that are possibly located next to areas of strategic importance.

Debates about the security aspects of Russian property purchases in Finland, and the location next to objects of strategic importance in particular, are met with mixed feelings by Russian owners. Some owners have expressed serious concerns:

> We are especially concerned by the discussions about property expropriation from present owners. We believe in sound Finnish reasoning. We would not like to lose our beloved second home.
>
> (Female, St. Petersburg, Russia, summer 2017)

Uncertainty and confusion are another line of reaction to these debates. Russian owners are puzzled by the notion of strategic objects and what kind of impact they might have on their second-home ownership:

> I would understand the nature of these debates if the land would be next to strategic objects. There aren't any objects around me, but you know, one can make up anything. Even a mobile phone station can be categorized as a strategic object. Of course, I am concerned, but I hope that Finland is a civilized country and if some decision is made, they wouldn't take my property away, but offer some compensation.
>
> (Male, St. Petersburg, Russia, summer 2017)

Social discourses and local attitudes towards Russian second-home ownership contribute to the outward image of the destination, which might influence the purchasing behavior of potential buyers and the attitudes of current owners. Social debates and legislative changes do not always lead to specific changes in traveler behavior and mobility practices. They are, however, a part of the social resilience as they shape among others the sense of security and institutional trust (Hannonen 2016; Lipkina 2013; Prokkola 2019). In fact, despite the debate and legislative change, Russian second-home owners maintain confidence about their property rights in Finland, showing a stable institutional trust. The social debate over Russian second-home ownership reflects a long-term social change that has been enshrined in law. The current data does not allow an estimation as to whether and how this change would affect the resilience of future property owners. However, at present, owners demonstrate confidence and trust in Finland's property rights, showing stable perspectives and value sets irrespective the debate and the legislative change, placing this free choice on the open end of the continuum.

Concluding discussion

This chapter applies the concept of social border resilience to understand the coping strategies of Russian second-home owners within the changing cross-border context since the year 2013. As societies evolve, their mobilities and communication practices transform, leading to new mobility regulations and practices. Thus, resilience is part and parcel of this process. Following earlier elaborations on resilience, tourism and borders (Cheer and Lew 2017; Humbert and Joseph 2019; Prokkola 2019; Saarinen and Gill 2018; Wandji 2019), this chapter looked into the responses of Russian second-home owners to mobility turbulences. The mobility dynamic across the Finnish-Russian border has been characterized by a range of changes that necessitated *fast* and *slow* responses in behavior and practices.

Border regulations and mobility policies that have been implemented to orchestrate and structure cross-border mobilities and flows became, paradoxically, turbulences *within* mobilities that disorder multiple mobilities connected to second-home ownership. The chapter has discussed mobility restrictions on certain professional groups under the sanction regime, changes in the visa regime, societal debate, bank policy and regulation of foreign assets. The border has become a central infrastructural and institutional marker in mobility modifications that caused predominantly inflexible solutions. The various changes evoke different degrees of adaptive capacity. Adaptive strategies for each change create a continuum of social border resilience practices with open and closed possibilities for resilience at the opposite ends of the continuum. The open-closed resilience continuum offers an analytical tool to examine the adaptive capacity of human agency in social border resilience. While in this chapter, the analysis has focused on the resilient practices of Russian second-home owners, the open-closed resilience continuum can be applied and adapted to other border and mobility research contexts as well.

In the case of Russian second-home mobility to Finland, the border plays a multidimensional role. Delimitating the external border of the EU, it translates political and geopolitical changes that directly affect the residents and visitors to the border area. The border also encompasses other supranational (e.g. Schengen area) and national regulations that impose additional sets of requirements on border crossers. In this regard, regulations on international mobility and legislative changes impose specific practices and behaviors without leaving space for resilient solutions. This underscores the importance of power and high politics in the discussion of social border resilience.

Moving and living between two states, Russian second-home owners develop new personalized solutions to meet the changed circumstances on both sides. In the case of navigating bank payments in Finland, resilience has turned out to be a "survival strategy." The variety of payment methods is, however, an example of creative solutions to a problem. On the other hand, the local social network has become an important asset in adapting to and meeting the restrictive regulations. This demonstrates that social border resilience is, indeed, a situational and contextual project. Focusing on the role of human agency in resilience in a cross-border mobility context, the chapter supports the notion of co-construction of resilience through practices and responses (Andersen and Prokkola 2022; Wandji 2019). The resilience continuum in turn structures these responses and defines the limits of adaptive capacities and resources, as the examination of social border resilience through mobility perspectives in the given border context points out. While the case takes place in the Finnish border region, a number of changes, such as the declaration of foreign assets or societal discourse, extend the presence of the border into the everyday. This emphasizes the relevance of studying the social border resilience and resilient practices of borderlanders also outside the border vicinity.

References

Amore, A., Prayag, G., and Hall, C.M., 2018. Conceptualising destination resilience from a multilevel perspective. *Tourism Review International*, 22, 235–250.

Andersen, D.J. and Prokkola, E-K. 2022. Introduction: embedding borderlands resilience. *In*: D.J. Andersen and E-K. Prokkola, eds. *Borderlands resilience: transitions, adaptation, and resistance at borders*. London: Routledge, 1–18.

Atorus., 2020. Association of tour operators. Statistics on Russians' travels abroad. Available from: www.atorus.ru/news/press-centre/new/50475.html

Brown, K., 2016. *Resilience, development and global change*. Abingdon: Routledge.

Cheer, J.M. and Lew, A.A., 2017. Understanding tourism resilience: adapting to social, political and economic change. *In:* J.M. Cheer and A.A. Lew, eds. *Tourism, resilience and sustainability: adapting to social, political and economic change*. Oxon: Routledge, 3–17.

Cresswell, T. and Martin, C., 2012. On turbulence: entanglements of disorder and order on a Devon beach. *Tijdschrift voor Economische en Sociale Geografie*, 103 (5), 516–529.

Draft Law N624870–6, 2014. Федеральный закон "О внесении изменений в отдельные законодательные акты Российской Федерации в целях введения запрета определённым должностным лицам иметь недвижимое имущество за границей" (Federal Law on amendments to legislative regulations to prohibit some Russian professionals to have property abroad). Available from: www.duma.gov.ru

Draft Law N920563–6, 2015. Федеральный закон О внесении изменений в статью 15 Федерального закона "О порядке выезда из Российской Федерации и въезда в Российскую Федерацию" и статьи 14, 17 Федерального закона "О службе в органах внутренних дел Российской Федерации и внесении изменений в отдельные законодательные акты Российской Федерации" (Federal Law on amendments to the law on exit from and entry to the Russian federation and the law on service in agencies of internal affairs of the Russian Federation). Available from: https://sozd.duma.gov.ru/bill/920563-6

Eskelinen, H. and Jukarainen, P., 2000. New crossings at different borders: Finland. *Journal of Borderlands Studies*, 15 (1), 255–279.

Federal Agency for Tourism, 2020. Statistics on mutual visits of citizens of the Russian Federation and other foreign countries. Available from: http://tourism.gov.ru/contents/statistika/

Federal Law FZ-140, 2015. Федеральный закон "О добровольном декларировании физическими лицами активов и счетов (вкладов) в банках и о внесении изменений в отдельные законодательные акты Российской Федерации" от 08.06.2015 N 140-ФЗ (Voluntary Declaration of Assets and Accounts (Deposits) in Banks). Available from: www.consultant.ru/document/cons_doc_LAW_180745/

Ferdoush, M.A., 2022. Stateless' yet resilient: refusal, disruption and movement along the border of Bangladesh and India. *In*: D.J. Andersen and E-K. Prokkola, eds. *Borderlands resilience: transitions, adaptation, and resistance at borders*. London: Routledge, 106–118.

Gurova, O. and Ratilainen, S., 2016. From shuttle traders to middleclass consumers: Russian tourists in Finnish newspaper discourse between the years 1990 and 2014. *Scandinavian Journal of Hospitality and Tourism*, 16 (Suppl. 1), 51–65.

Hall, C.M., 2018. Resilience theory and tourism. *In*: J. Saarinen and A. Gill, eds. *Resilient destinations and tourism: governance strategies in the transition towards sustainability in tourism*. London: Routledge, 34–47.

Hannonen, O., 2016. *Peace and quiet beyond the border: the trans-border mobility of Russian second home owners in Finland* (Publications of the University of Eastern Finland, Dissertations in Social Sciences and Business Studies, no 118). Tampere: Juvenes Print.

Hannonen, O., 2020. Strategic objective? contemporary discourse on Russian second home ownership in Finland. *In:* L. Lundmark, D. Carson and M. Eimermann, eds. *Dipping in to the north – living, working and travelling in sparsely populated areas*. Palgrave Macmillan, 311–332.

Hannonen, O. and Prokkola, E-K., in press. Physical access and perceived constraints: borders as barriers to travel mobilities and tourism development. *In*: D.J. Timothy and A. Gelbman, eds. *Routledge handbook of tourism and borders*. London: Routledge.

Hannonen, O., Tuulentie, S., and Pitkänen, K., 2015. Borders and second home tourism: Norwegian and Russian second home owners in Finnish border areas. *Journal of Borderlands Studies*, 30 (1), 53–67.

HE 253/2018vp. 2018. Hallituksen esitys eduskunnalle kansallisen turvallisuuden huomioon ottamista alueiden käytössä ja kiinteistönomistuksissa koskevaksi lainsäädännöksi (Governmental proposal to the parliament on taking into account security aspects in legislation on area use and property ownership). Available from: www.eduskunta.fi

Humbert, C. and Joseph, J., 2019. Introduction: the politics of resilience: problematising current approaches. *Resilience*, 7 (3), 215–223.

Lipkina, O., 2013. Motives for Russian second home ownership in Finland. *Scandinavian Journal of Hospitality and Tourism*, 13 (4), 299–316.

Lipkina, O. and Hall, C.M., 2013. Russian second home owners in Eastern Finland: involvement in the local community. *In:* M. Janoschka and H. Haas, eds. *Contested spatialities, lifestyle migration and residential tourism*. Oxon: Routledge, 158–173.

Notice., 2015. Предупреждение для российских граждан, выезжающих за границу [Notice for Russian citizens travelling abroad]. *The Ministry of Foreign Affairs of the Russian Federation*, 22 May. Available from: www.mid.ru/foreign_policy/news/-/asset_publisher/cKNonkJE02Bw/content/id/1305570

Order N154., 2018. Приказ Федеральной службы охраны РФ от 2 марта 2018 г. N 154.

"О внесении изменений в приказ ФСО России от 14 апреля 2015 г. N 168 "О сроках принятия мер по отчуждению имущества, право собственности на которое зарегистрировано за пределами Российской Федерации (Order of the federal guard service on amendments to terms of accepting measures concerning the real estate alienation that is registered outside the Russian Federation). Available from: https://rulaws.ru/acts/Prikaz-FSO-Rossii-ot-02.03.2018-N-154/

Order N168., 2015. Приказ Федеральной службы охраны РФ от 14 апреля 2015 г. N 168 "О сроках принятия мер по отчуждению имущества, право собственности на которое зарегистрировано за пределами Российской Федерации (Order of the federal guard service on terms of accepting measures concerning the real estate alienation that is registered outside the Russian Federation). Available from: https://garant.ru

Paasi, A., 1995. The social construction of peripherality: the case of Finland and the Finnish-Russian border area. *In:* H. Eskelinen and F. Snickars, eds. *Competitive European peripheries.* Berlin: Springer, 235–258.

Paasi, A., 1999. Boundaries as social practice and discourse: the Finnish-Russian border. *Regional Studies*, 33 (7), 669–680.

Pitkänen, K., 2011. Contested cottage landscapes: host perspective to the increase of foreign second home ownership in Finland 1990–2008. *Fennia*, 189, 43–59.

Pitkänen, K., Hannonen, O., Toso, S., Gallent, N., Hamiduddin, I., Halseth, G., Hall, C.M., Müller, D.K., Treivish, A., and Nefedova, T., 2020. Second homes during corona – safe or unsafe haven and for whom? reflections from researchers around the world. *Finnish Journal of Tourism Studies*, 16 (2), 20–39.

Prokkola, E-K., 2019. Border-regional resilience in EU internal and external border areas in Finland. *European Planning Studies*, 27 (8), 1587–1606.

PTK 16/2014. Preliminary debate, minutes. Available from: www.eduskunta.fi

Regulation (EC) No 610/2013 of the European Parliament and of the Council of 26 June 2013. Available from: http://data.europa.eu/eli/reg/2013/610/oj

Salovaara, O., 2016. Venäläisasiakkaiden suomalaisia pankkitilejä irtisanottu yllättäen [Russians' bank accounts have been unexpectedly terminated]. *Taloussanomat*, 25 November. Available from: www.is.fi/taloussanomat/art-2000004880008.html

Saarinen, J. and Gill, A., 2018. *Resilient destinations and tourism: governance strategies in the transition towards sustainability in tourism.* London: Routledge.

SKT 144/2018 vp. Oral question hour on foreign property purchases in Finland, minutes. Available from: https://www.eduskunta.fi/FI/vaski/Kasittelytiedot Valtiopaivaasia/Sivut/SKT_144+2018.aspx

Smętkowski, M., Németh, S., and Eskelinen, H., 2016. Cross-border shopping at the EU's eastern edge – the cases of Finnish-Russian and Polish-Ukrainian border regions. *Europa Regional*, 24 (1–2): 50–64.

Vihavainen, S. and Laitinen, S., 2018. Hallitus haluaa rajoittaa maakauppoja [The government wants to restrict land purchases]. *Helsingin Sanomat*, 26 August, A8.

Virtasalo, S. and Snieders, H., 2013. European Union – new rules for short-stays in Schengen Area. A publication for global mobility and tax professionals by KMPG's international executive service practices, December 9, 2013, No 2013–163, 1–4.

Wandji, D., 2019. Rethinking the time and space of resilience beyond the west: an example of the post-colonial border. *Resilience*, 7 (3), 288–303.

7 "Stateless" yet resilient

Refusal, disruption and movement along the border of Bangladesh and India

Md Azmeary Ferdoush

Introduction

On a sunny and warm Thursday morning, I sat down with Nur Alam in a tea shop in the Bangladeshi border village of Dasiar Chhara. Nur Alam is an unassuming 29-year-old who could not contain his excitement while sharing his role in the enclave exchange movement that went on for almost two decades across the borders of India and Bangladesh. A few years ago, the village that we were sitting in used to be part of India. It was one of the 162 enclaves that dotted the northern border between the two countries from 1947 until their exchange in 2015. Following the exchange, Dasiar Chhara was transferred to Bangladesh, and all the residents who opted to remain were accepted as Bangladeshi citizens after decades of an essentially stateless existence. Nur Alam did not strike as a revolutionary, but he had actively been part of a coordinated enclave exchange movement. When the movement began, it seemed like an impossible dream as well as a risky job. How could a group of de facto stateless people, who often became victims of sovereign violence, lead a political movement (Jones 2009a; Shewly 2013a)?

Nur Alam explained,

> People used to make fun of us. They used to say that we have gone crazy. India is never going to sign a deal that will make them to give up land [to Bangladesh]. But we were adamant. We carried on and gradually people started to support us. And look now, where we are! I am talking to you sitting here on the soil of Bangladesh!

This chapter presents a narrative of a transnational movement led by a group of stateless people living in the former border enclaves of Bangladesh and India. Enclaves were pieces of one sovereign state's land completely surrounded by the other; i.e., Indian enclaves were surrounded by Bangladesh and vice versa (van Schendel 2002). People living in these enclaves were, therefore, Indians living inside Bangladesh and Bangladeshis living inside India. Since these enclaves were completely cut off from their home state and an international border ran between them, the people living in the enclaves essentially ended up being landlocked in the host state, which did not formally recognize them as its citizens. They were

DOI: 10.4324/9781003131328-9

not protected by law, nor were they eligible to access any state services such as health and education, which effectively turned them into a "stateless" population (Jones 2010; Shewly 2013a; Cons 2016; Ferdoush and Jones 2018). Yet, they were successful in leading a coordinated transnational movement by forming the India Bangladesh Enclave Exchange Coordination Committee (IBEECC hereafter) in 1994. The IBEECC and their movement not only demonstrate that living in precarious situations for a prolonged time does not necessarily turn people into passive victims of violence but, at the same time, complicates our understanding of resilience and resistance along borders.

The enclaves of Bangladesh and India have their roots in the pre-colonial era. They were created as a result of a treaty between the Mughal ruler and the Cooch Behar King in 1713 (van Schendel 2002; Whyte 2002; Jones 2010). Both the rival parties were in frequent conflict and controlled a portion of lands inside each other's territories. They agreed to retain those lands for tax collection purposes in the treaty of 1713 and hence gave birth to the enclaves that would later come to trouble the imaginations of state- and nation-making in post-colonial South Asia (Ferdoush 2019a). These enclaves became an international issue for the first time in 1947 when colonial India was divided between the two sovereign states of Pakistan and India. Pakistani enclaves ended up on the other side of the border and vice versa. At the beginning, residents in these enclaves could move with ease to their home state, but as days passed by and bordering became hardened due to numerous factors including, but not limited to, erection of border fences and deployment of border guards, they ended up being locked inside the host state. Pakistan and India agreed to exchange their enclaves in 1958, but, for several reasons, it was never executed. East Pakistan, which hosted all the enclaves, gained independence through a bloody war and became the sovereign state of Bangladesh in 1971. As Bangladesh was now hosting Indian enclaves, it signed another treaty in 1974 with India to exchange them. The treaty is known as the 1974 Land Boundary Agreement (LBA hereafter). Both parties agreed to give up their claim on the enclaves across the border and merge the hosting enclaves as regular state territories. They further agreed that the enclave residents would be given an option to choose the state where they wanted to become citizens (Ferdoush 2019b). The treaty took another 41 years to be executed, leaving enclave residents in limbo. Since neither the host state recognized them nor the home state maintained any connection, they eventually ended up being a "stateless" population. Such a status marginalized the enclave people both politically and economically but could not turn them into passive recipients of power and violence. They became resilient through day-to-day "survival strategies" and by "refusing" to abide by the rules of the sovereign (Jones 2012; Shewly 2016). One of the most telling aspects of their "refusal" was the enclave exchange movement and the creation of IBEECC.

Drawing on the movement of the former enclave residents, in this chapter, I present a narrative of a resilient borderland people who often refused to abide by the rule of the sovereign. In so doing, I problematize the idea of resistance by demonstrating that enclave people's movement and other survival strategies are best understood as *acts of refusal* instead of binary opposition to power and dominance. While resilience itself is a "notoriously slippery concept to pin down" (Grove 2018, p. 5) and more than 40 different usages of the term can be traced (Tierney 2014), my purpose here is not to engage with what resilience is. Instead, I follow the editors' view that "borderlands resilience is about people's many ways of 'making sense' of what is 'a life worth living'" (Andersen and Prokkola 2022, p. 7). I pay particular attention to the "making sense" and "a life worth living" aspects of resilience by focusing on the enclave exchange movement.

The findings are based on field research in June and July of 2015 and between June 2017 to May 2018 as the former enclave residents settled into their new-found homes. During this time, I stayed and moved around eight of the biggest enclaves inside Bangladesh in three northern districts of Rangpur division. I conducted a total of 102 in-depth interviews and six focus group discussions with numerous groups, including the former enclave residents, regular Bangladeshi citizens, administrative officers, local leaders, and political figures (see for detail Ferdoush 2020). In the discussion that follows, I first shed light on the enclave exchange movement led by the IBEECC. Then I unpack the concept of *acts of refusal* and position it within the scholarship of the geography of resistance. Finally, in conclusion, I discuss the potential of *acts of refusal* in forwarding our understanding of resilience along borders.

The enclave exchange movement

The unique situation of being stuck in a country that they did not belong to put the former enclave residents in a precarious position. They lacked protection by law and were often made victims of sovereign violence. Scholars have fruitfully applied the Agambenian (Agamben 1998, 2005) framework of violence and abandonment to understand the former enclaves as "container of bare lives" (Shewly 2013a), as "spaces of exception" (Jones 2009b) and as "sensitive space" (Cons 2016). Such works demonstrate that though the former enclave residents often became *homines sacri* whose lives could be spared without warranting any repercussion, they were not merely passive victims of violence. They came up with numerous tactics to survive by evading the sovereign through formal and informal arrangements. Such tactics included, but were not limited to, arranging a fake identity card, using an address of a relative from the host country, marrying someone in the host country or buying properties (Jones 2012; Shewly 2015; Cons 2016; Ferdoush 2021). However, these were mostly daily activities of survival adopted by individuals, not a coordinated movement connecting enclaves across the border. Yet, the IBEECC and their movement have hardly gained attention from academics except partially by Hosna Shewly (2015).

Dipak Sengupta, who founded the IBEECC and initiated the enclave exchange movement, was not an enclave resident. Instead, he was a regular Indian citizen and a four-time elected Member of the Legislative Assembly (MLA) in Cooch Behar under the State of West Bengal. The IBEECC was formed in 1994 in Dinhata constituency in Cooch Behar district of India but failed to gain traction at its earlier phase. However, after Dipak expired, his son Diptiman Sengupta took over and coordinated the movement. In the Bangladesh side, it was Gulam Mostafa who pioneered the exchange movement as the General Secretary of the Bangladesh chapter of IBEECC. Mostafa was a life-long resident of Dasiar Chhara enclave, and I found him to be the most visionary and well-spoken among all the leaders I interviewed. In his late 40s, Mostofa "managed" to get a university degree from one of the prominent institutions of Bangladesh and founded a girl's school in Dasiar Chhara in 2015 after the enclave was exchanged. Mostafa came to know about Dipak Sengupta's initiative and thought of doing the same in Bangladesh. Thus, by the early 2000s, he started connecting with local leaders around the neighboring enclaves in Bangladesh and from India too. Eventually, Mostofa got a letter from Dipak inviting him to attend a meeting in Dinhata in 2004. So, he started for Dinhata with a few of his fellow enclave residents. Not all of them had passports, and they hoped to cross the border by playing[1] their identity of an enclave dweller (for identifications and play, see also Andersen 2022).

Those who are familiar with the Bengal borderland are well aware that crossing the border frequently happens with informal arrangements (Jones 2009a, 2012). Such crossings often depend on making deals with brokers who would then find a way to evade the border guard using cross-border networks or simply finding a way to dodge the system. However, the most significant aspect of these crossings is that these are not performed as active resistance to the state power. Instead, people *refuse* to abide by the rules of the sovereign because they often cannot afford to follow the official procedures of border crossings. Sometimes livelihood and survival depend upon such crossings, and occasionally precarious situations like that of the enclave residents compel them to do so as they are not even considered eligible for a passport, in the first place (Jones 2012; Shewly 2015). Therefore, refusal remains an integral part of survival and eventually of resilience along the border of Bangladesh and India.

Mostofa and his team went in and attended the meeting. In Mostofa's words,

> I went to the meeting and was given a chance to speak. So, I stood up and said that from today all the enclave residents are brothers. It does not matter whether we are Hindus or Muslims, there is only one identity we hold, and that is, we are all *chhit bashi* (enclave dwellers). On my way back, Dipak *kaka* (uncle) gave me a copy of the 74 LBA and explained the whole treaty to me. I came back and initiated to form a chapter of the committee in Bangladesh, and we finally had a committee in 2006. After the death of Dipak *kaka*, his son Diptiman took over, and the movement became more intense from 2009 [explanation added].

Organizing and leading such a movement was a challenging task not only in itself but also because the enclave dwellers lacked identity as rights-bearing citizens. Since they were not recognized by Bangladesh as citizens, they lacked access to resources and power to negotiate with the government officials and state mechanisms. It did not take them long to come up with a solution to this problem. The solution was to select Moinul Islam as the president of the Bangladesh chapter of IBEECC. Moinul was not only a Bangladeshi citizen but also, the local Chairman[2] with political connections and resources. As Mostofa puts it:

> We decided to make him [Moinul] the president because he was very sympathetic to us. Also, he was the Chairman, and he was from Bangladesh. We needed someone who had a voice and who could talk for us.

The former enclave residents had a clear agenda, which proves that they could "speak" for themselves, but at the same time, they were aware of their "inferior rank" and the inability attached to that rank to reach to and negotiate with the state elites (Guha and Spivak 1988; Spivak 1988). Thus, the crucial question for them was, "how to make their voices heard?" Selecting Moinul as the President, on the one hand, ensured that their voices would be heard and, on the other, provided them access to valuable networks they lacked. This also brings us back to the question of resilience posed by Andersen and Prokkola (2022). The forming of IBEECC and the movement itself demonstrate the resilience of a de facto stateless population who continuously worked their ways through precariousness by "making sense" of what is "a life worth living." In other words, they refused to remain silent in the face of the inactive role by the state, and through numerous acts of refusal, including the exchange movement, they honed resilience.

After forming the Bangladesh Chapter of the IBEECC and setting up the office in Dasiar Chhara, they concentrated more on organizing and connecting the rest of the enclaves and their residents. Mostofa, Nur Alam, Moinul and other members of the committee started traveling to the neighboring enclaves and eventually visited all the enclaves in Bangladesh. It took them almost three and a half years to finally bring all the enclaves under the umbrella of IBEECC. Moinul described the process:

> We started to connect with other enclaves of Panchagar, Lalmonirhat, Kurigram, and Nilphamari.[3] Most of the time, Mostofa and I used to ride a motorcycle to visit other enclaves. We formed different district committees and enclave committees. It took us three and a half years to form a unit for each of the enclaves and tie them together.

They formed three district-level committees in Panchagar, Kurigram and Lalmonirhat districts. As Nilphamari district hosted only four enclaves, it was tagged with Panchagar. Each district had unit committees. Kurigram, Lalmonirhat, Panchagar and Nilphamari respectively had 6, 12, 25, and 4 such unit committees. They used to coordinate between and among them via mobile telephone. The IBEECC performed numerous activities, including submitting memo reminder to different government offices, organizing local meetings, engaging with the field level government offices, coordinating with the Indian chapter and so forth.

Altaf Hossain, the founding president of Dasiar Chhara unit of the IBEECC, described their activities:

> We started, first, by giving letters to the government of Bangladesh and India. But nothing happened. No one cared. We formed a committee of 60 members in Dasiar Chhara. Then the Awami League gov. came into power [in Bangladesh] and the Congress [in India] also came into power at the same time. We saw it as a huge opportunity. So, I called for a meeting in Dasiar Chhara on 21st February of 2009, the International Mother Language Day. We hoisted the Bangladeshi flag in Dasiar Chhara on that day. . . . Then we went to the UNO office and gave him a memo reminder.

Nur Alam added,

> We came up with a slogan. The slogan was *"74'er chukti, chitbashir mukti"* (Treaty of 74, enclave dwellers' freedom).

The IBEECC also maintained a regular connection with the local administration, realizing that it was highly important to maintain a network with the state and state actors from a precarious position like theirs. But the journey was not always as smooth as it sounds. The IBEECC leaders often had to work out serious allegations and opposition both from other enclave residents and state apparatuses. Being the president and the general secretary of the IBEECC, Moinul and Mostofa faced the worst. Moinul shared one such incident of facing the SP (Superintendent of the Police – the highest-ranked police officer in charge of a district in Bangladesh) of Kurigram district. Moinul was sent a letter to his office from the SP that alleged him and Mostofa of conspiracies against the state, and both were summoned to the SP's office. Moinul said,

> Mostofa and I were summoned by the SP. We were very nervous, but we had no option except to see him. With a beating nervous heart, we entered his office. There were a number of journalists already in his office. However, the SP asked the journalists to wait outside and talked with us with the ASP present. We explained to him in detail about the inhumane conditions of the enclave dwellers and what we were trying to do [through IBEECC]. We said, look, we are not doing anything against the state; we just want the treaty that was signed in 1974 to be executed. He was well convinced. He told us that the information he had was totally opposite of what we told him. Then he suggested us to carry on. Also, he wished us good luck. After that, the local police never interrupted us anymore, if not helped.

Both Awami League and the Indian National Congress Party won the general election and formed a government respectively in Bangladesh and India in 2009. This was the same combination that signed the 1974 LBA. Awami League is historically known to have a friendly relation with India, especially with the Indian

Congress Party (Ferdoush 2019a). The combination fashioned much optimism and further intensified the exchange movement. Right after being elected as the Prime Minister of Bangladesh, Sheikh Hasina paid a state visit to India in January 2010 and expressed her desire to solve all the boundary disputes in the spirit of the 1974 Agreement (Ferdoush and Jones 2018). The state-level dialogue brought much enthusiasm among the former enclave residents and fueled the IBEECC.

After the state visit of Sheikh Hasina in 2010, bilateral issues gained pace. Various disputed border concerns were solved, undemarcated borders were marked within a year of the visit, and both states agreed on an exchange of the enclaves soon (Shewly 2013b). Such developments brought much hope, and the former enclave dwellers called for a *Moha Somabesh* (Grand Meeting) in Rangpur, the northern division of Bangladesh that accommodated all the enclave hosting districts. H. M. Ershad, a former president of Bangladesh, was the chief guest, while Diptiman Sengupta from India, coordinator of the IBEECC, was the special guest. Residents from both Bangladeshi and Indian enclaves attended the meeting on February 15, 2011.

Bangladesh and India conducted a joint survey in all the former enclaves for the first time from July 14 to 17 of 2011. The census not only counted the population but also asked for their choices of citizenship. The survey was conducted as a preliminary step for exchanging the enclaves and brought an unprecedented sense of relief among almost all the enclave residents as they could, for the first time, see the exchange to be eminent. Soon after the headcount, the Indian Prime Minister Manmohan Singh was scheduled to visit Bangladesh from September 6 to 7, 2011, and during this visit, he was supposed to sign the enclave exchange treaty. However, at the last moment, the Chief Minister of West Bengal, Mamata Bannaerji, opposed the treaty for numerous reasons. Thus, no treaty was signed; instead, it resulted in a land boundary protocol without a definite time period for the exchange (for details see Cons 2014; Shewly 2013b). The lack of a fixed timeframe in the protocol sparked a wave of dissatisfaction among the enclave dwellers on both sides, and they came up with numerous activities to protest the decision. Among these, the most serious and impactful was a hunger strike demanding a specific date for the exchange and the ratification of the 1974 LBA. Moinul and Mostofa, being the top two leaders of the IBEECC of Bangladesh chapter, shared their experiences of the hunger strike.

Mostofa: Finally, we started the strike in the Putimari enclave.[4] We decided this to be the venue because a highway runs over it that connects Panchagar district with Dhaka [capital city of Bangladesh] and the rest of the country. Diptiman and others were also doing the same on that side [India] [explanation added].

Moinul: It was almost four weeks. We were still holding our positions, but no one seemed to care. We decided to block the highway and stop the traffic movement. A traffic jam formed very soon. Then the SP of Panchagar came to see us and requested us to withdraw the road blockage and ensured us that the government is keeping a close eye on the strike. We withdrew the road blockage at his request and instead blocked the rail track. We would not move until the government sent its representative to listen to us and promise a specific deadline. On the 26th day, several local MPs (Member of the Parliament) and leaders came to meet us and expressed their support for us. The next morning one of the editors of a leading national daily came to see our strike. Also, the Prime Minister sent her representative to us, and she promised to press the issue with India. She [the representative] came with a truck of juice and food to end our hunger strike.

The strike went on from March 18 until April 11, 2012, and only ended with a promise from the government of Bangladesh to press the issue with India (Shewly 2015; Ferdoush and Jones 2018). The hunger strike was a great success both because it gained unprecedented attention in India and Bangladesh and was a strong demonstration of their resilience. In the subsequent years, though enclave issues got overshadowed by other geopolitical events, the IBEEC continued its activities and finally celebrated the execution of the exchange of enclaves at midnight of July 31, 2015. Enclaves did not exist after August of 2015, and neither did the IBEECC, but the office of the Bangladesh chapter of IBEECC[5] was still carrying the legacy of a resilient borderland population even in 2018 when I visited them.[6]

Acts of refusal

How, then, are we to comprehend the enclave exchange movement, and how does such a take contribute to our understanding of resilience along borders? In answering this question, I start first by situating the enclaves and their movement within two predominantly applied frameworks – state of exception and dominance-resistance. In doing so, I argue that none of these are fully capable of capturing the nuances of the movement and offer the *acts of refusal* as an alternative.

As has been mentioned earlier, the abandonment of the former enclave people has been largely understood using the Agambenian framework of state of exception and violence (Jones 2009a, 2009b; Shewly 2013a). Such frameworks simply view that the extraterritoriality of the enclaves created conditions for the exceptional rule. As a result, until the enclaves were exchanged, they remained spaces of exception, where the rule of law was suspended, and the enclave people's bodies were given extrajudicial status who could be killed without consequences. This framework leaves hardly any space for resistance, or in other words, assumes that in spaces of exception the sovereign rule leaves no crack or fissure through which (everyday) resistance can emerge. Edkins and Pin-Fat thus rightly contend that theorization of sovereign power in such a way nullifies any meaningful resistance since those can be neutralized with the use of exception (Edkins and Pin-Fat 2004; Jones 2012).

On the other axis of such a view is the dominance/resistance scholarship, which claims resistance to be present everywhere within the framework of power (Sharp et al. 2000; Hughes 2020). In this view, every action from people at the receiving end of power is an act of resistance against the dominant/powerful, ranging from simply not following social norms to movement against the authority. The most well known of such works is James Scott's formulation of almost any noncompliance to be resistance in the absence of a full-fledged revolution (Scott 1985). In geography, Sharp et al. offers a similar perspective as they have it:

> No moment of domination, in whatever form, is completely free of relations of resistance, and likewise no moment of resistance, in whatever form, is entirely segregated from relations of domination: the one is always present in the constitution of the other.
>
> (Sharp et al. 2000, p. 20)

In their view, resistance could encompass acts ranging from "breaking wind when the king goes by" to "violent action" (2000, p. 3). The limitation of adopting such an all-encompassing view is that it makes the analysis of resistance "increasingly meaningless" (Jones 2012, p. 686). At the same time, it "dilutes the purchase of the term" (Hughes 2020, p. 1148).

Therefore, I suggest that viewing the (cross-border) movement as an *act of refusal* allows us to productively situate *refusal* in analyzing the resilience of a borderland population who neither wanted to change the existing power relations through their movement nor challenged the sovereign rule. Instead, they simply refused to accept the life conditions that they thought were not "worth living." Acts of refusal are thus expressed acts and actions from an individual or groups whose aim is not to override the existing order but to simply refuse to follow the rules that do not align with their perceived conditions of lives. In the current instance, the IBEECC and the exchange movement organized by them is an act of refusal from the former enclave residents as they denied living in limbo without a foreseeable solution to their situations through the ratification of the LBA. Acts of refusal can definitely "disrupt the ordering logic on which the state relies" but do not aim to alter them (Jones 2012, p. 687). To elaborate on the significance of acts of refusal, I shed light on three major characteristics in this chapter. First, acts of refusal are neither resistance, nor do they happen without the knowledge of the agent of the sovereign. The movement of the IBEECC, the *moha somabesh*, Moinul and Mostofa being summoned by the SP of Kurigram and, above all, the month-long hunger strike all demonstrate that the state and state agents were fully aware of their activities. In fact, some of those actions were actively supported by numerous state agents. As Shewly also found, none of these "actions in the host country's territory was conducted without authorization of local administration" (Shewly 2015, p. 21). Therefore, in this instance, the sovereign decided not to use its exception and violence on the enclave people, although it could have done at a moment's notice without any consequences. The IBEECC could operate, and the enclave exchange movement continued only because the sovereign allowed it. Therefore, this is a clear indication that there is room for refusal within the sovereign exception.

Second, *acts of refusal* do not aim to change or overturn the existing power relations and/or the status quo. Therefore, these are best understood not as resistance but as refusal or situated practices. Drawing on Cindi Katz, I suggest that the enclave exchange movement is best understood as reworking of power relations through formal and informal arrangements that made them more resilient to survive within a state of exception (Katz 2004). These are acts and actions of people who consciously chose to deny the sovereign authority either simply to survive and/or to increase their wellbeing. Therefore, the telling aspect of such acts is that they are performed on a case to case basis, as opposed to a routinized sub/un-conscious reproduction (Giddens 1984). In discussing the nature of such refusal, Jones has it, "People accept that the state is there and a categorical order has been imposed, but they do not necessarily accept those categories" (Jones 2012, p. 697).

Finally, by definition, *acts of refusal* create spaces of refusal or take place in spaces where power relations are continuously reworked. Scholars have pointed out that borderlands, camps and "sensitive spaces" are productive grounds which give rise to acts of refusal on a regular basis in numerous forms (Jones 2012; Cons 2016). However, drawing on the instance of the IBEECC and their activities, I further contend that acts of refusal do not only occur in such spaces but also give rise to spaces of refusal in the making. The month-long hunger strike and the *moha somabesh* by the IBEECC effectively turned regular state-spaces into spaces for staging refusal. The highway or the rail tracks are major state infrastructures that were used as space for staging the strike by obstructing the regular flow of business. The *moha somabesh* was organized at the heart of the city of Rangpur where people of the marginal background came together in numbers reevaluating what is a "life worth living" by refusing to accept the existing arrangements. Through their acts, they turned those spaces into spaces of refusal, even if temporarily. Therefore, acts of refusal demonstrate that borderland resilience is not only about direct resistance to the dominant power, nor are they exclusively concerned about changing the existing status quo. Resilience develops through the everyday practices of survival and ontological reflection of individual actors who constantly renegotiate their position within power relations.

Conclusion

The former enclave residents of Bangladesh and India lived in limbo for generations, waiting to be accepted as regular citizens by these states. The waiting was finally over in 2015 when the hosting states exchanged and merged their enclaves with regular state territories. At the same time, enclave residents were given the option to choose a state of citizenship. In this chapter, I have focused on their experiences of living in the period of the impasse when they led a cross-border exchange movement. Drawing on the coordinated enclave exchange movement, I have demonstrated that even their marginal conditions could not turn them into passive recipients of sovereign violence. Instead, they actively refused to abide by the sovereign rules. Their acts and actions of finding a way around and through the sovereign rule are understood as acts of refusal, which I contend allowed them an "ability to hang on to a sense of hope that gives meaning and order to suffering in life and help articulate a coherent narrative to link the future to the past and present" (Southwick et al. 2014, p. 10). As the pinnacle of their refusal, I have offered a narrative of the IBEECC and numerous tactics that the movement, consisting of former enclave residents, carefully adopted so that it would not trouble the sovereign in a manner to warrant the use of the exception. Therefore, neither with the purpose nor with the means, the enclave exchange movement was a resistance movement. It is better captured as acts of refusal. Such a take on refusal, I contend, enables a bottom-up approach in comprehending the resilience of borderland population in the ways identified next.

First, resilience can be productively identified and analyzed along borderlands by paying careful attention "to the context and institutional structures" (Andersen and Prokkola 2022, p. 6). Second, exclusion by the state does not necessarily result in *bare lives* who are stripped of all kinds of agencies. Instead, as the enclave residents demonstrated, the past experiences offer a resource that increases resilience. Third, the resilience of people living in precarious conditions is often warranted by their will to survive. Such survival instincts result in creative acts to "hang on" to the situation with the hope that gives meaning to their sufferings. Finally, viewing refusal as part of resilience could enable a productive dialogue without losing the nuances of daily life to the all-encompassing state of exception on the one hand, and not diluting the academic purchase of the term by embracing every act as resistance, on the other.

Notes

1 Until the 1990s, enclave residents used numerous informal arrangements for crossing the border. The most common of these was getting an enclave identity card from a local political leader. The Indian BSF were also aware of this and would let the enclave residents cross the border with such identity documents. However, since the 1990s, with hardening of the borders and bordering, these informal arrangements were stopped. Mostafa and his team hoped to use their identity as enclave residents to cross the border but were stopped by the BSF. Then, Dipak Sengupta had to intervene and make arrangements for Mostafa and his team to cross the border informally.
2 Chairman is the elected leader by popular vote at a sub-district level in Bangladesh. They hold an office for five years and work side-by-side with administrative structures of the state in planning and implementing numerous programs.
3 All the 111 enclaves were hosted by these four districts in Bangladesh.
4 This is a former enclave situated under Boda Upazila in Panchagar District of Bangladesh. The reason they chose it was that a national highway that connects Panchagar district to the rest of the country runs over this enclave.
5 Since the IBEECC ran a coordinated movement across the border of Bangladesh and India, both sides were commonly known as chapters, i.e., the India Chapter of IBEECC and the Bangladesh Chapter. Although IBEECC did not exist after the

exchange, they kept the office in Dasiar Chhara as an example of their sufferings, movement and resilience.

6 The Bangladeshi enclave of Dahagram-Angorpota was not exchanged as originally agreed in the 1974 LBA because Bangladesh gave up its partial claim on another Indian enclave (for detail, see Cons 2016; Whyte 2002).

References

Agamben, G., 1998. *Homo sacer: sovereign power and bare life*. Stanford, CA: Stanford University Press.

Agamben, G., 2005. *State of exception*. Chicago, IL: University of Chicago Press.

Andersen, D.J. 2022. Line-practice as resilience strategy: the Istrian experience. *In*: D.J. Andersen and E-K. Prokkola, eds. *Borderlands resilience: transitions, adaptation and resistance at borders*. London: Routledge, 166–181.

Andersen, D.J. and Prokkola, E-K., 2022. Introduction: embedding borderlands resilience. *In*: D.J. Andersen and E-K. Prokkola, eds. *Borderlands resilience: transitions, adaptation, and resistance at borders*. London: Routledge, 1–18.

Cons, J., 2014. Impasse and opportunity: reframing postcolonial territory at the India-Bangladesh border. *South Asia Multidisciplinary Academic Journal*, 10.

Cons, J., 2016. *Sensitive space: fragmented territory at the India-Bangladesh border*. Seattle, WA: University of Washington Press.

Edkins, J. and Pin-Fat, V., 2004. Introduction: life, power, resistance. *In*: J. Edkins, V. Pin-Fat, and M. Shapiro, eds. *Sovereign lives: power in global politics*. London: Routledge, 1–21.

Ferdoush, M.A., 2019a. Symbolic spaces: nationalism and compromise in the former border enclaves of Bangladesh and India. *Area*, 51 (4), 763–770.

Ferdoush, M.A., 2019b. Acts of belonging: the choice of citizenship in the former border enclaves of Bangladesh and India. *Political Geography*, 70, 83–91.

Ferdoush, M.A., 2020. Navigating the 'field': reflexivity, uncertainties, and negotiation along the border of Bangladesh and India. *Ethnography*. Epub ahead of print 12 July. DOI: 10.1177/1466138120937040.

Ferdoush, M.A., 2021. Sovereign atonement: (non)citizenship, territory, and state-making in post-colonial South Asia. *Antipode*, 53, 546–566.

Ferdoush, M.A. and Jones, R., 2018. The decision to move: post-exchange experiences in the former Bangladesh-India border enclaves. *In*: A. Horstmann, M. Saxer, and A. Rippa, eds. *Routledge handbook of Asian borderlands*. London: Routledge, 255–265.

Giddens, A., 1984. *The constitution of society: Outline of the theory of structuration*. Cambridge: Polity Press.

Grove, K., 2018. *Resilience*. London: Routledge.

Guha, R. and Spivak, G.C., eds., 1988. *Selected subaltern studies*. Oxford and New York: Oxford University Press.

Hughes, S.M., 2020. On resistance in human geography. *Progress in Human Geography*, 44 (6), 1141–1160.

Jones, R., 2009a. Agents of exception: border security and the marginalization of Muslims in India. *Environment and Planning D: Society and Space*, 27 (5), 879–897.

Jones, R., 2009b. Sovereignty and statelessness in the border enclaves of India and Bangladesh. *Political Geography*, 28 (6), 373–381.

Jones, R., 2010. The border enclaves of India and Bangladesh: the forgotten lands. *In:* A.C. Diener and J. Hagen, eds. *Borderlines and borderlands: political oddities at the edge of the nation-state.* Lanham, MD: Rowman & Littlefield, 15–32.

Jones, R., 2012. Spaces of refusal: rethinking sovereign power and resistance at the border. *Annals of the Association of American Geographers,* 102 (3), 685–699.

Katz, C., 2004. *Growing up global: economic restructuring and children's lives.* Minneapolis: University of Minnesota Press.

Scott, J., 1985. *Weapons of the weak.* Princeton, NJ: Princeton University Press.

Sharp, J.P., Routledge, P., Philo, C., and Paddinson, R., eds., 2000. *Entanglements of power: geographies of domination/resistance.* London: Routledge.

Shewly, H.J., 2013a. Abandoned spaces and bare life in the enclaves of the India – Bangladesh border. *Political Geography,* 32, 23–31.

Shewly, H.J., 2013b. Sixty six years saga of Bengal boundary making: a historical expose of Bangladesh-India border. *BIISS,* 34 (3), 205–219.

Shewly, H.J., 2015. Citizenship, abandonment and resistance in the India – Bangladesh borderland. *Geoforum,* 67, 14–23.

Shewly, H.J., 2016. Survival mobilities: tactics, legality and mobility of undocumented borderland citizens in India and Bangladesh. *Mobilities,* 11 (3), 464–484.

Southwick, S.M., Bonanno, G.A., Masten, A.S., Panter-Brick, C., and Yehuda, R., 2014. Resilience definitions, theory, and challenges: interdisciplinary perspectives. *European Journal of Psychotraumatology,* 5 (1), 25338.

Spivak, G.C., 1988. Can the subaltern speak? *In:* C. Nelson and L. Grossberg, eds. *Marxism and the interpretation of culture.* Basingstoke: Macmillan Education.

Tierney, K., 2014. *The social roots of risk: reproducing disasters, promoting resilience.* Stanford, CA: Stanford University Press.

van Schendel, W., 2002. Stateless in South Asia: the making of the India-Bangladesh enclaves. *The Journal of Asian Studies,* 61 (1), 115–147.

Whyte, B.R., 2002. *Waiting for the Esquimo: an historical and documentary study of the cooch behar enclaves of India and Bangladesh.* Melbourne, Australia: School of Anthropology, Geography and Environmental Science, University of Melbourne.

Part 3

Making time

Identity-formation and historical memory as resilience

8 Schleswig

From a land-in-between to a national borderland

Steen Bo Frandsen

Introduction

In a contemporary context, the concept of borderlands is most often connected to areas with a presence of nation-state borders. Understood from a historical perspective, however, this would not necessarily be the case. The concept of borderlands is particularly useful when discussing limitations to national historiographies because it opens towards an understanding of continuities, flows and contacts independent from the institutions of nation-state borders of a later date. In recent years, European historians have become more interested in bringing a borderlands concept into the analysis of European history, where borderlands define regions where empires and composite states met (O'Reilly 2018). Following Philipp Ther, viewed historically, borderlands might more precisely be characterized as "lands-in-between," thereby stressing ambiguities (Ther 2013), mixed cultures and overlaps that characterized most border regions before they were accommodated by nation-state imaginaries with their particular understand of borders and reproduced as integrated parts of monochromatic nation-states on the European map.

Contemporary borderlands were often such "lands-in-between." People living there would have no idea that their region would soon be divided by state borders, and thus end up being called "border regions." Such processes should not be ignored when discussing borderland resilience and borderland identities in a contemporary context because it is in these processes that regions acquired the prefix "border" through which they are characterized today. People living in what is now border regions had to accommodate to this new situation, they often had to resist and to deal with competing neighbors or national ideologies, and these processes have marked and formed border regions in specific ways (see also Ridanpää 2022; Andersen 2022).

In this chapter, borderland resilience is discussed in relation to processes of changing from a land-in-between to a national borderland. Schleswig constitutes an example. For centuries the region formed a part of the Oldenburg composite monarchy (Østergård 2014). From the middle of the 19th century Schleswig became the scene of one of Europe's first conflicts over a national border, and

DOI: 10.4324/9781003131328-11

since 1920, the region has been divided between Denmark and Germany. Historians almost exclusively discussed the developments from the perspective of a change of nationality, a perspective revealing a nationalist and state-center dominated approach. Following the overall understanding of borderland resilience, this chapter will, by contrast, understand the developments from the perspective of the border (region) in order to highlight a resilience dimension in the regional context (cf. Andersen and Prokkola 2022).

A historical approach to borderland resilience

Composite states contained different territories with various status including a variety of languages, laws and cultural characteristics (Elliott 1992; Gustafsson 1998). Regions situated in the zones of transition, i.e. between the composite states were often more ambivalent and generally characterized by multiple cultural influences. They would have complex histories, particular privileges, mixed populations and identities, all of which was formed by their geographical position at the edges of the composite states. Often, they would be objects of desire and war, and end up as prey or loss. There was no given outcome with respect to their future nation-state affiliation. Even if these contested regions nowadays find themselves inside a particular nation-state, their histories would derive from competing national narratives and interpretations (Ther 2013; Readman et al. 2014; Frank and Hadler 2015).

Nation builders and national historians placed borders and border regions in the periphery of the map and the nation-building process, thereby ascribing them a prominent role in the national narrative. The fact of being contested areas has secured them a lasting place in collective memories, yet, never in terms of their own histories being in any way central to describing the nation-state. Nation-states promoted a dominant view from the center, placing the border regions firmly inside their respective national narratives. When we discuss borderland resilience and identity in the following, we will turn the perspective and see the developments from the border (Rumford 2012; Andersen and Prokkola 2022, p. 000).

The unsettled and transitional character of these regions has been ignored for long within the dominant historiographical nation-state approach to borders and the regions defined by them. However, when considering borderland resilience, it is necessary to recognize the endurance of historic regions. The sheer existence of historic regions with names and notions of traditions and memories are signs of histories of resilience and identity. State and nation-building processes might, together with national historiography, have done the utmost to erase them, yet they often have much longer trajectories than national historians claim. In certain cases, changing borders has not led to the loss of identification with a region that might no longer be the same as it was in the past. According to Philipp Ther, here even national identities "appear volatile and context-dependent" compared to regions. In Ther's work, this is illustrated in what he calls "intermediary spaces" like Upper-Silesia and Alsace (Ther 2013, 496).

The efforts to dominate the border regions in the process of bringing intermediary spaces under the control of the nation-state touch upon the issue of center and periphery. Border regions might have strong interests in keeping up relations to centers outside the emerging nation-state for economic or cultural reasons, providing prominent motives for resilience (Rokkan and Urwin 1983). Nation-state borders would threaten the ambiguity and the advantages of being in between. Regions about to be included in nation-building processes therefore had reasons to hold onto known structures and confront ideas of change and bordering with a resilient attitude. Border regions would not always consider themselves to be peripheral and vulnerable areas. Their efforts to resist becoming simple peripheries and losing a regional freedom to act would be an argument for holding on to known structures and habits.

Despite Eugen Weber's critical discussion of nation-building processes in his book *Peasants into Frenchmen*, it is argued that his center-periphery structure is too simple. It hardly concerns opposing influences and does therefore not pay much attention to the characteristics of border regions involved in complex processes of nationalization (Weber 1979). Research has challenged the center-dominated view with examples from nation-state peripheries like the Pyrenees (Sahlins 1989) or the Flemish region of northwestern France, thus demonstrating how competing centers and influences from outside the nation-state could play an important role in forming regional identities able to position themselves somewhere in between the nation-states (Laven and Baycroft 2008).

Generally, national historiographies would deal with border regions from an outside perspective and place them firmly within a nation-state narrative. Consequently, their take on borderland resilience and identities would be included into their own narratives. When local actors opposed changes and threats coming from outside their region, these actions were understood as support for a national rather than a regional cause. In the case of Schleswig, resilience is overwhelmingly interpreted from a national and not a regional perspective. Danish historians have almost exclusively focused on the actions of the Danish side, seeing resilience as a general opposition to and reaction against German influence. Schleswigians, however, held a long record for rejecting Danish initiatives that were considered a threat to local culture and practice. Also, regionalists or Schleswig-Holsteinian separatists were simply considered as foes.

This chapter discusses borderland resilience with the border region perspective as point of departure. "Seeing from the border" (Rumford 2012) includes the wish to retain traditions to preserve a borderland characteristic of multiple cultures and influences, of vernaculars and in-between realities opposing the idea of disappearing in a homogeneous nation-state, and the literature and sources used in this chapter therefore also gather examples and indications of borderland resilience that would mostly not be included in the mainstream interpretation.

The end of the composite state and the birth of a borderland

Since medieval times, Schleswig was an independent duchy. Located in between the Danish kingdom and the duchy of Holstein, it was influenced by mixed cultures many languages and inner territorial partitions. The duchies reflected the composite structure of the German Empire, also underlining the transitional and mixed character of the duchies where the Danish king was the elected duke of the Danish fief of Schleswig and the German fief of Holstein (Porskrog-Rasmussen 2006; Østergård 2011; Frandsen 2016).

The loyalty of the Schleswigians towards the Danish king was never questioned, but the duchy was well known for local privileges and strong local patriotism: The Frisians living in the western parts and on the islands were proud of their local culture and keen to defend their rights, and the peninsula of Eiderstedt, the area of Angeln, the wealthy farmland of Loit to the north and the island of Fehmarn had strong local identities. Something similar could be found in Flensburg and other towns. These sentiments were particularly widespread among conservatives and the economic elites throughout the region – peasants as well as merchants with interests in markets and trading contacts outside Schleswig.

Fertile ground for resilient cultures reacting against changes threatening the status quo were thus widespread in the region. When state authorities attempted to modernize worship and introduce a rationalist reform of the Church in the late 18th century, the plan was met with strong opposition throughout the duchy. Typically, protests were directed against those public servants in favor of changes, yet leaving the king uncriticized (Henningsen 2016). Conflicts did not question the unity of the kingdom or the independence of Schleswig until the national conflict raised the question of the future of the duchy in the late 1830s – the question that would eventually bring an end to the Oldenburg monarchy.

Schleswig became the main battlefield during the fight between Danish nationalists and the German-oriented separatists of Schleswig-Holstein. The Danes wanted to annex the duchy and make it an integral part of the kingdom. Their opponents started out as regionalists fighting for the independence of the duchies within the Oldenburg monarchy but then drifted more and more towards separatism, later even demanding to be included in a German state. Hence, Schleswig began to emerge as a border region because of the aspirations to end its position as a transitional region. Demands to either include it in a Danish or a German state, or even divide it, introduced a border-discourse that would expose Schleswig's location in a center-periphery dimension.

Schleswig makes a strong case of how a land-in-between became a borderland in the nation-state sense. In these processes, the borderland failed to live up to the expectations of the conflicting parties, being neither Danish nor German. A breeding ground was thereby fertilized for resilience opposing the efforts to submit and uniform the region according to national ideas. Becoming a borderland changed the perspective and the room for action for the Schleswigians, clearly identifying great risks of being swallowed by nation-states and trying to find strategies to secure their regional identity and independence.

The Schleswigians had profited from the infrastructural position of their region binding the different parts of the monarchy together, and the demand that Schleswig should choose between north and south confronted it with a completely new center-periphery dimension. Being equidistant from the competing centers, Copenhagen and Hamburg, Schleswig would be in focus of the rivalry. Some Schleswigians might have had preferences as to which center to choose, but generally there was growing uncertainty, the Schleswigians having successfully organized themselves in a way that would profit as much as possible from the proximity to both centers, an outcome of being located comfortably in between.

Resilience may not in itself be a question of numbers, but as national historiography rather ignores the existence of people thinking differently in national matters, it supports the argument. It remains unclear how many people subscribed to these ideas, but they were undoubtedly more common than later historians would admit. All petitions rejecting a division brought forward before the outbreak of violence in 1848 received more support than the nationalist petitions demanding a one-sided Danish or German solution to the problem. Nationalism was not backed unanimously as it was later claimed, and even the Danish national movement had to consider regional identities. This was acknowledged in the most successful of all petitions before the war of 1848: The Schleswigians were characterized not as Danes but as "in-betweeners" (Schultz Hansen 1997, 2016).

Jens Wulff (1774–1858), a wealthy merchant of lace from Bredebro in Western Schleswig, personified economic success, and a conservative worldview made him a supporter of the composite state. His diaries portray him as an outspoken opponent of nationalist movements with their propaganda of hate and antagonism, clearly preferring the world he knew: a peaceful, mixed society loyal to the king. He was tolerant vis-à-vis different people and local traditions commonly found in his beloved region, and he rejected the seemingly crazy idea that people should be divided according to their language (Iversen 1954–1956).

Wulff was not alone in his wish to avoid radical changes or in his rejection of a definition of nationality based on language. Occasionally, voices were raised in the public, criticizing the nationalist ideas. One of the leading regional newspapers, *Lyna* from northern Schleswigian Haderslev, published ten protest addresses with 525 signatures mostly from peasants in 1843–1844. They condemned the Danish infiltration and expressed the wish to go on living in peace and harmony with all their fellow Schleswigians. The Schleswigians should neither be divided from their German neighbors nor annexed by the Danes (Schultz Hansen 1997).

The protesters underlined their regional identity by a condescending attitude towards their neighbors to the north yet wanting to keep close ties to the kingdom. Like Wulff, they could not accept that language should define nationality. Schleswigians spoke at least three languages but formed only "one nationality" as they expressed it, thereby applying nationalist language to their cause. In another regional petition, more than 1,500 persons declared they were and intended to remain Schleswigians. From these protests, Schleswig emerged as an independent region between the Danish kingdom and the duchy of Holstein, claiming a specific regional identity founded upon old privileges and institutions and rejecting a nexus between language and identity (Schultz Hansen 1997). Similar convictions were dominating in Flensburg, the biggest and wealthiest town of the region. The successful merchants were closely connected to the monarchy by privileges and trade interests and, at the same time, some of them had interests in keeping up or even expanding the contacts to the booming center of Hamburg.

Opinions were divided, which illustrated how the balancing and compromising of a land-in-between came under still more pressure. In Flensburg, cultures and languages met. Most people, including those favoring a Danish orientation, spoke German, something which had not previously been an issue, as loyalty towards the king was decisive. The elite would agree with Wulff regarding the question of language and identity, not qualifying Flensburg as Danish nor German, and a predominant feeling existed that it was impossible to choose between Germany and Scandinavia. A choice in either direction would threaten the wealth and the prospects, and consequently the merchants tried their best to stay outside the conflict and remain a Schleswigian city (Henningsen 2009). Not only Flensburgians but most Schleswigians felt secure within the multi-ethnic and composite Oldenburg monarchy as it provided room for intermediate identities reflecting the regional reality. Before the First Schleswigian War (1848–1850), the efforts to remain Schleswigian and avoid becoming a victim of the nationalists' demands for dividing borders could be understood as a resilient position. The wish to save the mixed character of a province without a clear national orientation represented a bid for "a good life," as it is formulated in the diaries of Jens Wulff.

Under growing pressure from Danish and German nationalists, this region of transition turned more and more into a contested border region. Many regretted this, but their compromising conviction seemed rather weak and colorless confronted with the dramatic language, aggressive claims and violent threats of the nationalists. Any idea of a Schleswigian regionality or even "nationality" turned out to be a chimera. The antagonism between Danish and German outgrew the regional discourse, and in the end the nationalists succeeded in imposing the idea of a national border that the Schleswigians had tried so hard to avoid. The national border – regardless of it being included in a Danish or a German state, or even dividing it between them – has dominated the historical narrative of Schleswig ever since. It created the false idea of Schleswig having been a national borderland since times immemorial.

The Schleswigian War marked a turning point. The Schleswig-Holsteinian uprising followed by military interventions by the Danish and German federal armies turned the region into a bloody battlefield, polarizing the region and resulting in a much harder division among the Schleswigians. It became almost impossible to argue for in-between solutions leaving aside questions of saving the Schleswigian identity against the pressure from outside. The national borderland was dominated by resistance against one of the nationalities respectively and ambitions to force the region to become either Danish or German. Confronted with becoming borderlanders in a nation-state, the Schleswigians were now forced to choose between their cultural, political and economic dependence something that was probably more often than not articulated against one of the conflicting parties rather than being a real choice in favor of one of them.

The Germanification of Angeln

The translation of regional resilience into the national discourse can be seen in the most prominent case of the so-called "change of nationality" that took place in Angeln, an area situated just southeast of Flensburg. Praised for its natural beauty and rich agrarian culture, Angeln became the showcase for the nationalist claim of ongoing Germanification of the Schleswigians. German language had for a long time been the prestigious language compared to the vernaculars, and since the beginning of the 19th century Germanification was strengthened by the growing importance of economic orientations towards Hamburg. Danish nationalists now became convinced that a change of language – and therefore nationality – was happening before their very eyes. Combined with suggestive but imprecise language maps, it created the apocalyptic vision of a continuous advance of German language and culture on what was understood to be "old Danish ground" appealing to Danes in the kingdom (Figure 8.1, Adriansen 1990).

Nationalists effectively instrumentalized these events to show a steady Germanification and dramatic language shift. Observing how the Danes would change into Germans, this was not a very precise analysis, and the reality was much more complex. The nationalists never operated with distinctions between Danes living in the kingdom and those living in Schleswig, thereby ignoring influences of mixed culture in the identity of Schleswigians who also considered themselves Danish – misunderstandings, illustrated by the one-dimensional concept of language. As described earlier, Schleswigians rejected a language-defined identity, knowing from practice in their everyday life advantages of using different language (thereto, see also Andersen 2022). These languages were not standardized, and people living in Angeln spoke vernaculars much closer to each other than the talk about Danish vs. German suggested. On the one hand, this made it easier to communicate, yet it also accommodated a shift of language (Wolbersen 2016). In Angeln, it was possible to identify a growing use of the German "platt," a result not of a campaign, but for practical reasons because Angeln was a wealthy farmland with connections out of the region. To local farmers, higher social prestige connected with German, just like most products of German culture, were considered "en vogue."

Angeln came to the attention of nationalists as a zone of transition between the Danish-dominated northern and the German-dominated southern part of the duchy. Angeln therefore seemed decisive for winning control over Schleswig. After the war of 1848–1850, the Danish authorities decided to confront the situation by actively promoting the official Danish language here as well as in other areas of Central Schleswig. During the 1850s, the local reaction against this initiative constituted an example of resilience very much in line with earlier reactions against efforts to change what the regional population considered to be their

Figure 8.1 On the language map of Schleswig presented by the Danish historian C.F. Allen in 1857, the border of the Danish language was pushed much too far to the south. This led to a widespread but biased belief in the strength of Danish culture in the region, and it was used by nationalists in their demands.

Source: Adriansen 1990.

culture and way of life, reminding of earlier protests against the introduction of a rationalist worship, but now the conflict was nationalized. The Danish government expelled German-speaking vicars of Schleswig-Holsteinian conviction and substituted them with imported clericals from Denmark. When it was announced that Danish would be the language spoken in church every other Sunday, the local population stayed away (Jensen 1844).

The language policy and the repressive actions against opponents rebounded after the Schleswig-Holsteinians had lost the war. Loyal German-speaking citizens that had contributed to Flensburg's image as a pro-Danish city reacted bitterly disappointed and repeated their rejection of a link between language and identity (Grænseforeningen 1955). The language policy provoked critical reactions from abroad, but worst of all it consolidated an anti-Danish conviction that would prevent the Danes from winning back the mixed areas once and for all. Angeln became a stronghold in the fight against the Danes. During the war of 1864, Danish priests and civil servants were driven out by the raging local population (Linvald 1920), and as the Danish army retreated from its southernmost line of defense hundreds of Schleswigian soldiers deserted rather than followed the defeated army back to Denmark (Borberg 1938).

Borderland and nation-state

The Second Schleswigian War of 1864 was another gamechanger in the borderland. The composite state was gone, and the population had to adjust to a new situation. After defeating Denmark, Prussia seemingly put an end to the discussion about the future of the region by annexing Schleswig. The national conflict almost resulted in a division of Schleswig, the northern part being overwhelmingly Danish while the center and the south had clear German sympathies.

The context of resilience changed and could no longer be separated from the national struggle. Many lines pointed towards the past though. Schleswig was still considered a problematic region, and the Prussians would have to deal with regional identities too. The Prussian conquest was highly unpopular with most Schleswigians. They had never dreamt of becoming Prussians, and the German-minded population demonstrated the wish for an independent Schleswig-Holsteinian state. This slowly began to change after the Unification of Germany of 1871, but many Schleswigians found it hard to accept the "Prussian way." Instead of becoming a free regional state, Schleswig and Holstein became provinces of Prussia. Among the issues were the spreading of High German at the cost of local vernaculars, and the introduction of a foreign school system with a curriculum that did not respect regional traditions. Still, the German Schleswigians celebrated that their region had been saved from partition. They did not think about a return to the Danish state.

Among the Danes the situation was very different. They rejected the Prussian dominance, and their dislike of the new masters only grew with the German Unification. A stubborn fight against the politics of Germanization began, and this further strengthened the orientation towards the Danish language and culture. The goal was to become a part of Denmark, and even if the prospects were bleak, the fight was mobilizing the population. Lost in their own arrogance, the Prussians repeated earlier mistakes of their adversaries. The forced Germanization of the overwhelmingly Danish-minded parts of the region turned out to be counterproductive, and a nation-building process also took place among the Danish. Their fight was supported by the kin-state. The fate of the duchy and the resilience of their fellow countrymen inspired the nation-building process in the Danish state north of the border.

All Schleswigians – Danes or Germans – kept alive the idea of a region but with not much left of the positions that had characterized a Schleswigian conviction in the early days of national conflict in the composite state. The German side clung to the dogma of an indivisible region without paying attention to the Danish character of the north; some Danes also continued dreaming of a united Danish Schleswig, but after 1864 more and more of them realized that the best they could get would be national division.

Schleswig divided

Following the decision of the Paris Peace Conference in 1919, two plebiscites were held in Schleswig in early 1920 to draw a border according to the principle of national self-determination. The voters were only given the simple choice between Danish or German. The results were clear, Schleswig was divided with its northern part annexed by Denmark. The new border reflected the division which had in effect already been the outcome of the national conflict in the middle of the 19th century. The results of the plebiscites were clear, but the chosen border created two national minorities, and disappointment remained on both sides.

The plebiscite offered no middle way, no Schleswig to vote for, and thus no common ground left outside the national antagonism. This was different in plebiscites in other borderlands like Silesia (Ther 2013), but in Schleswig decades of nationalization had established two distinct national cultures. Older ideas of Schleswig might still exist on both sides, and a regional consciousness remained, but Danish and German Schleswigians were no longer able to unite on these grounds. The new national border changed the meaning of resilience once again because now the Schleswigians had to learn how to live and deal with a dividing border. The scene was set for nation-building and integration processes that would deepen the differences inside the region and draw the two parts into the respective nation-state.

Although a Schleswigian region no longer existed and the two parts would soon end up being peripheries, none of them would become like the rest of their respective nation-states. The border would become important in the daily lives and the development of the Schleswigians. The German minority in Denmark and the Danish and the Frisian minorities in Germany would recall the former Schleswig and keep variation alive. The Danish and the German minorities, the biggest losers of the plebiscites, would also create difficulties after 1920.

The Schleswigian borderland remained unsettled. Economic troubles following the World War made it very difficult to impose "the national order", and considerable mistrust towards the borderlanders remained. German authorities had feared an even greater territorial loss. They took up the suitable argument of an imminent threat from the Danes to claim state-subsidized programs to build up institutions and infrastructure in the border region. Explicitly, they mentioned the need to build up a resilience among the less convinced German-minded Schleswigians towards the Danes by showing the potentials of the German republic (Frandsen 1994, pp. 132–133).

More surprising was perhaps the fact that the Danes were confronted with a much more difficult situation than expected. Northern Schleswig returning "home" was considered "justice done," yet disappointment with the new reality was notable, and the integration of the province proved troublesome. The population of the Schleswigian borderland was influenced by a long regional tradition and by many aspects of the Prussian system it had experienced for almost 60 years. The majority did not regret becoming Danish, yet change was not necessarily seen as progress, and the wish to mark a regional difference was apparent. A surprisingly high number of German-minded clergymen won the elections held to oust them, a sign not of pro-German sentiments, but of the tradition of regional resilience against decisions or actions considered to be in conflict with the wish to keep the congregation united across the national divide. This could also be understood as protection against too eager a Danish effort to reform the Church of Schleswig (Schwarz-Lausten 2020).

The clear results of the plebiscites blurred the fact that people on both sides felt uncomfortable with answering yes or no to a complicated question like national identity. The border did not prevent people from moving to the other side not least because the new border divided many families. One of those crossing the border was the painter Emil Nolde, who was born on the Danish side but settled down on the German side. A large majority of the population were to some extent mobilized by the national movements, and among the elite hardly anyone would remain outside the national organization. Yet, some groups remained out of reach, maybe also because the highly polarized atmosphere did not provide much room for the voices of those who would see themselves between the poles.

The existence of people indifferent to the national question was not denied. In meticulously gathered statistics, the authorities searched for those that had not been convinced by the national propaganda. They caused much anxiety because they were considered to form an unreliable potential that could be mobilized by the other side. On the Danish side, they would be known with a specific pejorative term that signified a mixed color ("de blakkede"). Their numbers were relatively small, they were especially found among the lower social strata of the population, and, typically, they would not read the (national) press. Quite a lot of them would turn up in the suspicious category of "mixed marriages". In the nationalized borderland, those who did not see themselves as Danish or German acted resilient in the sense that they did not submit to the prescribed categories and expected behavior (Jepsen 2004); they just tried to get along. Borderland people who more or less openly challenged the national order always risked exclusion. One example of this being Germans south of the border who after the Second World War received food aid from the Danish side and who would be called "Speckdänen," meaning that they had been bought by the Danes.

As the nation-state represented the end of history, the border was expected to be definitive. Hence, within the nation-state there would be no need or reasons for resilience against the state. Probably under influence of difficult experiences in the first decade after the Danish annexation, the former national activist Adolph Svensson described his fellow Schleswigians, identifying three main elements of their nature: A deep inherited Danish nationality, a pronounced conservatism with respect to habits and world view, and a general opposition towards those in power typical for borderlanders. While the first would facilitate nation building, the second only briefly interfere with it, the fundamental characteristic of opposing the rulers might still be an issue in the future, if the borderlanders were not treated respectfully (Svensson 1930).

After the first somewhat bumpy experiences with reunification, a certain anxiety grew among the Schleswigians. They were not as nationally conformed as the nationalists had expected. As borderlanders, they were used to "one storm after another", leading to more or less open suspicion towards all things coming from the outside (Svensson 1930). A borderlander himself, Svensson was convinced that the border would influence the people and their actions and that this would not change in the nation-state.

The border of 1920 had created national minorities on both sides. They were unhappy with their placement in their respective states and kept up the hope to join their kin state. It was easier for the Germans to claim they were treated unfairly in the border-drawing process, but between the World Wars and during the German occupation of Denmark (1940–1945) their position became seriously damaged because of their collaboration with the National Socialists. Interestingly, the German government took no initiatives to redraw the border during the occupation of Denmark, and even if this did not mean anything definitive given the National Socialists' general disregard of borders, it reflects the importance of the decision of 1920. After 1945, Danish nationalists demanded a border revision, a momentary interest for Denmark south of the border, which reintroduced old Danish national ideas of winning back the region.

Instead, the situation calmed down. Relations between Denmark and the Federal Republic developed in a more peaceful way as a sort of parallel lives without much incitement to develop contacts or cross-border activities. This was demonstrated with the declarations signed in Bonn and Copenhagen in 1955 that finally produced an understanding in relation to the minorities. The division between the nationalities was hardly questioned, and regional resilience – a Schleswigian identity – no longer played a role in culture or politics. In 1997, the historian Hans Schultz Hansen concluded that it was doubtful if the concept of Schleswig could work as a common idea for regional institutions across the border, referring to the discussions about a Euroregion in the public debate at the time (Schultz Hansen 1997).

The Danish admission to the Common Market in 1973 was strongly supported in Northern Schleswig, but it did not mark the beginning of a significant change in the relations to the south. The emotional debate about Denmark's belated admission to the Schengen agreement in 2001 once more manifested how far the region had moved away from the past as a space of transition. Although the opposition to the open border was much stronger in the distant capital, no one in the region questioned the national border. In fact, the opening of the border after 2001 facilitated a change in the border region and contributed to reduce the obstacle it constituted to both sides. Growing contacts and not least the closer cooperation between the minorities could be understood as a cautious reawakening of borderland thinking (thereto Andersen and Winkler 2020).

Conclusion

In the process from being a land-in-between characterized by a mixed culture and multiple languages to becoming a borderland divided by nation-states, clearly demarcating their territory and demanding more homogeneous culture on both sides, Schleswig shows similarities with other European borderlands with a past in between composite states. Already as a land-in-between, a resilient behavior was a prominent and characteristic feature. The Schleswigians reacted when traditions and privileges came under pressure from the outside. An idea of a region with a mixed culture profiting from its position in between was articulated clearly. However, from the mid-19th century, it became still more difficult to retain a regional position due to conflicts between nationalists and separatists demanding clear choices. Resilience did not disappear, but it had to adapt to the new conditions itself as well.

The changes following the establishment of a national border dividing the region in 1920 can be put into perspective by another borderland of the Oldenburg composite state. From early on, Danish historians have compared 19th-century Schleswig with the traumatic loss of the region of Scania to Sweden two centuries before. A parallel was seen in what was defined as "a change of nationality," thereby ignoring that nationality would change over time and might have different meanings in different contexts. The perspective does not "see from the border" and thus ignores how both conflicts took place in a composite state-setting very different from the nation-state. Yet, when we consider how these two borderlands developed after the end of the composite states, the comparison offers interesting points for the discussion of borderland resilience and identity.

Historians disagree about the character of the Swedish annexation of Scania in 1658 (Frandsen and Johansen 2003; Gustafsson 2006; Jespersen 2010). At first, the Swedish king followed a rather cautious strategy to secure his new possessions, like how other composite states tackled such challenges. During the Scanian War (1675–1679) the Danish army returned to Scania in a bid to take back the lost province, but in the end the Danes were defeated and driven out. The war made the Swedish king change his mind, introducing a repressive strategy to secure lasting control over the region and make it clear that there would be no turning back. This widely successful strategy has been read as a "Swedification" (Fabricius 1906–1958). Other historians however have stressed the context of the composite states and found it much more relevant to talk about a "Scanification" (Sanders 2008).

Both arguments could be true, however. Scania remained a part of Sweden, the border to Denmark became a national border, and Scania could not be considered a region in between. Yet, Scania remained undivided. It became a part of Sweden, but arguably it did not stop being Scania. Today the region presents its own flag and demonstrates a regional pride, and Scanians like to stress that they are closer to the world outside than other parts of Sweden.

Scania was estranged from the rest of Denmark and the old capital of Copenhagen, but the Swedish capital remained very far away. Even before the annexation, the border advised by nature was rather the broad and trackless zone of forests to the north of Scania, separating it from Sweden. This is probably why it still makes sense to talk about a "Scanification" in a longer perspective and why the concept of resilience helps us to understand the longevity of local identities and feelings of being something apart. The potential for resilience and regional identity remained, and Scania would hardly ever have been able to claim its regionality if it had stayed within the centralist Danish kingdom with a dominant capital just around the corner. Wandji argues that the border community does not seek transformation and is resilient essentially in terms of adaptation as a form of continuity rather than change (Wandji 2019). This is arguably true in the case of Scania. The undivided region was in many respects too distant to become Swedish. In the longer perspective, resilience was not least the ability to retain a special identity. Since the building of the bridge across Øresund, Scania profiles itself against both the Swedish state and the ambitious neighbor across the water.

Returning to Schleswig, the situation was clearly very different. The conflict also had its origins in the composite state, Schleswig changing from a land-in-between to a region defined by the border drawings resulting from the national conflict. But it was never "Schleswigianized." On the contrary, the competing national ideologies defeated the old regionalism together with ideas of transitional space and a multilingual society as identity-defining characteristics. The national conflict divided the Schleswigians in two main groups, and they would not again come together with an idea of a common regionality, a common Schleswig. Hence, a regionality like in Scandia did not materialize. Having been nationalized during the wars of the 19th century, the divided groups would orientate themselves towards centers and political processes outside the region. In Scania, a strong borderland identity profited from the distant capital and a regional feeling of being close to the world outside Sweden. In Schleswig, the capitals were relatively distant and absent, and after the division both parts were too weak to play anything else than a symbolic role as a national border region.

Borderland resilience in a historical perspective has a lot to do with seeing from the border and being aware of a regional reality that differs from that of other regions in the nation-state as well as from the view from the center. Border regions thereby gain an active role in the process, and this makes it possible to problematize nation-state narratives. Borderland resilience, understood as a totality of traditions and experiences of living with and adapting to the border underlines that borderlands are dynamic places, defined by their histories and the border but in continuous movement and change vis-à-vis the border.

References

Adriansen, I., 1990. *Fædrelandet, folkeminderne og modersmålet*. Sønderborg: Museumsrådet.

Andersen, D.J., 2022. Line-practice as resilience strategy: the Istrian experience. *In:* D.J. Andersen and E-K. Prokkola, eds. *Borderlands resilience: transitions, adaptation, and resistance at borders*. London: Routledge, 166–181.

Andersen, D.J. and Prokkola, E-K., 2022. Introduction: embedding borderlands resilience. *In:* D.J. Andersen and E-K. Prokkola, eds. *Borderlands resilience: transitions, adaptation, and resistance at borders.* London: Routledge, 1–18.

Andersen, D.J. and Winkler, I., 2020. Grænsearbejde blandt grænsependlere (Special issue on Borders, Steen Bo Frandsen, ed.). *Oekonomi og Politik,* 2, 56–68.

Borberg, L., 1938. *I krig og kantonnement 1864.* København: Hagerup.

Elliott, T.H., 1992. A Europe of composite monarchies. *Past & Present,* 137, 48–71.

Fabricius, K., 1906–58. *Skaanes Overgang fra Danmark til Sverige, I-IV.* København: Nordisk Forlag.

Frandsen, K.-E., and Johansen, J.C.V., eds., 2003. *Da Østdanmark blev Sydsverige.* Højbjerg: Skippershoved.

Frandsen, S.B., 1994. *Dänemark. Der kleine Nachbar im Norden.* Darmstadt: Wissenschaftliche Buchgesellschaft.

Frandsen, S.B., 2016. Some reflections on Schleswig and Holstein as contested regions. *In:* M. Bregnsbö and V. Jensen, eds. *K Schleswig Holstein – contested region(s) through history.* Odense: UP SDU, 15–25.

Frank, T. and Hadler, F., eds., 2015. *Disputed territories and shares pasts. Overlapping national histories in modern Europe.* London: Palgrave MacMillan.

Grænseforeningen., 1955. *Flensborg Bys Historie,* vol. 2. København: Grænseforeningen.

Gustafsson, H., 1998. The conglomerate state: a perspective on state formation in early modern Europe. *Scandinavian Journal of History,* 23 (3–4), 189–212.

Gustafsson, H., 2006. Att testa gränser. *In:* H. Gustafsson and H. Sanders, eds. *Vid Gränsen. Integration och identiteter i det förnationella norden.* Göteborg: Makadam, 7–18.

Henningsen, L.N., ed., 2009. *Sydslesvigs danske historie.* Flensborg: Studieafdelingen ved Dansk Centralbibliotek for Sydslesvig.

Henningsen, L.N., 2016. *Værdikamp og folkeuro: bønder, præster og øvrighed i 1790'ernes Slesvig.* Aabenraa: Institut for Grænseregionsforskning.

Iversen, P.K., 1954–1956. Kniplingskræmmer Jens Wulffs Dagbog. *Sønderjyske Aarbøger,* 66, 60–130, 67 (1), 55–152, 67 (2), 184–227.

Jensen, H.N.A., 1844. *Angeln.* Geschichtlich und topographisch beschrieben (Neu bearbeitet und bis auf die Gegenwart fortgeführt von W Martensen und I Henningsen). Jul. Vergas Verlag und Druckerei: Schleswig 1922.

Jepsen, S.K., 2004. De blakkede. National indifference og neutralitet i Nordslesvig 1890–1940. *Sønderjyske Aarbøger,* 67–86.

Jespersen, L., 2010. Grænser og grænseproblematikker. *In:* A. Palm and H. Sanders, eds. *Flytande gränser. Dansk-svenska förbindelser efter 1658.* Göteborg: Makadam, 23–42.

Laven, D. and Baycroft, T., 2008. Border regions and identity. *European Review of History,* 15 (3), 255–275.

Linvald, A., 1920. Stemninger og Tilstande i Sønderjylland ved Krigens Udbrud 1864. *Danske Magasin,* 6R, III.

O'Reilly, W., 2018. Frederick Jackson Turner's frontier thesis, orientalism, and the Austrian Militärgrenze. *Journal of Austrian-American History,* 2 (1), 1–30.

Østergård, U., 2011. Schleswig and Holstein in Danish and German historiography. *In:* T. Frank and F. Hadler, eds. *Disputed territories and shared pasts: overlapping national histories in modern Europe.* London: Palgrave Macmillan, 200–223.

Østergård, U., 2014. Nation-building and nationalism in the Oldenburg empire. *In:* S. Berger and A. Miller, eds. *Nationalizing empires.* Budapest: Central European University Press, 461–509.

Porskrog-Rasmussen, C., 2006. The Duchy of Schleswig – political status and identities. *In:* H. Gustafsson and H. Sanders, eds. *Vid Gränsen. Integration och identiteter i det förnationella norden.* Göteborg: Makadam, 180–203.

Readman, P., Radding, C., and Bryant, C., 2014. *Borderlands in world history, 1700–1914.* London: Palgrave Macmillan.

Ridanpää, J., 2022. Borderlands, minority language revitalization and resilience thinking. *In:* D.J. Andersen and E-K. Prokkola, eds. *Borderlands resilience: transitions, adaptation, and resistance at borders.* London: Routledge, 137–151.

Rokkan, S. and Urwin, D.W., 1983. *Economy, territory, identity. Politics of west European peripheries.* New York: Sage Publications.

Rumford, C., 2012. Towards a multiperspectival study of borders. *Geopolitics*, 17 (4), 887–902.

Sahlins, P., 1989. *Boundaries: the making of France and Spain in the Pyrenees.* Berkeley, CA: University of Califormia Press.

Sanders, H., 2008. *Efter Roskildefreden 1658.* Göteborg: Makadam.

Schultz Hansen, H., 1997. Schleswigsche Identität in den 1840er Jahren – ein historischer Begriff wird wieder aktuell. *Grenzfriedenshefte*, 239–252.

Schultz Hansen, H., 2016. The influence of economic and social interests on the choice of nationality in Schleswig 1840–1848. *In:* M. Bregnsbo and K. Villads Jensen, eds. *Schleswig Holstein – contested region(s) through history.* Odense: UP SDU, 121–146.

Schwarz-Lausten, M., 2020. *Den kirkelige genforening 1920.* København: Kristeligt Dagblad.

Svensson, A., 1930. Folke-Sind mellem Skamling og Flensborg Fjord. *In:* A. Nordahl-Petersen, ed., *Turistforeningen for Danmark.* Aarbog: Sydøst-Jylland, 137–147.

Ther, P., 2013. Caught in between. Border regions in modern Europe. *In:* O. Bartov and E.D. Weitz, eds. *Shatterzone of empires: ethnicity, identity, and violence in the German, Habsburg, Russian, and Ottoman Borderlands.* Bloomington, IN: Indiana University Press, 485–502.

Wandji, G., 2019. Rethinking the time and space of resilience beyond the West: an example of the post-colonial border. *Resilience: International Policies, Practices and Discourses*, 7 (3), 288–303.

Weber, E., 1979. *Peasants into Frenchmen.* Stanford, CA: Stanford University Press.

Wolbersen, H., 2016. *Der Sprachwechsel in Angeln im 19. Jahrhundert.* Hamburg: Kovac Verlag.

9 Borderlands, minority language revitalization and resilience thinking

Juha Ridanpää

The creation of modern nation-states and their borders has played a key role in the process through which Meänkieli, a language spoken in northern Sweden, has turned into and been acknowledged as a minority language. The history of Meänkieli as a borderland language goes back to the beginning of the 19th century when, in the aftermath of the war of 1808–1809, the territory of Finland was separated from Sweden and annexed to Russia as the autonomous Grand Duchy of Finland. As a result, the new border, a line drawn along the Torne River, divided the historically, linguistically, economically and culturally integrated region of the Torne Valley between two states. The condition of being a linguistic minority began to develop into an issue of social "othering" in the late 19th century when ruthless "Swedification" policies entailed the exercising of powerful political pressure on marginal groups in order to integrate them linguistically and culturally into the modern nation-state.

Today Meänkieli is classified as an endangered language, and there is a perception of the group of Meänkieli speakers as passive, oppressed and harshly treated by the majority population. As has been the case with several subaltern languages, the feelings of shame remain embedded in the self-perception of the speakers. In case of minority groups and languages, resilience thinking refers to an alternative to common top-down language policies, a bottom-up approach, in which the decision-making concerning the revitalization of endangered languages is given to local groups from the hands of state governments implementing their political and socially unifying purposes (Bradley 2019). To understand how language resilience works, it is essential to understand the context and theoretical framework of colonization, marginalization and trauma (Fitzgerald 2017, p. 281). In the case of Meänkieli, the question of linguistic sovereignty is inherently connected to the negative connotations of Finnishness within Swedish society and the sense of shame that Meänkieli-speaking people feel for having Finnish roots. On the other hand, there is an alternate perception of Meänkieli speakers as a group which is, in spite of oppression, actively involved in several cultural projects that aim to protect their language and increase the esteem of a cross-border identity. In the studies of resilience linguistics, it has been particularly underscored how work aiming to revitalize endangered languages can function as a mechanism for

DOI: 10.4324/9781003131328-12

strengthening fragile minority groups and communities, a "bounce back" from a disturbance of socio-cultural marginalization.

This chapter discusses language resilience thinking with the focus on how the conflicting viewpoints over the socio-political status of minority languages may turn into simultaneous acceptance of uncertainty and hope for the better future. The specific attention is paid to the changing role of bordering, which in the case of the Meänkieli language and identity represents a symbolic marker for a shameful past. The main research question concerns how cultural activists involved in Meänkieli language revitalization re-narrativize their shifting identities by connecting resilience thinking together with the practices of active socio-cultural resistance. The study is based on group discussions conducted in northern Sweden during the fall of 2015 and the spring of 2016 with Meänkieli-speaking cultural activists.

Borders and minority language revitalization

In minority language studies, language loss has often been linked with the context of globalization, along with arguments such as how globalization has brought along a so-called "reversing language shift" (Fishman 1991). The threat of globalization is in connection with a widely employed juxtaposition in which endangered languages are associated with conservatism, as they are assumed to be spoken by a soon-passing generation, while younger generations stand for the ideal and values of modernism, liberalism and globalization (see for example Dorian 1994). On the other hand, the loss in linguistic diversity has also been explained being an impact followed by the development of centralized nation-states (Bradley 2019, p. 509). The condition of being linguistically marginalized effectively means a reduced sense of belonging to the state (Valentine and Skelton 2007). Both these viewpoints are justified within their own contexts, but what is crucial to emphasize here is how major a role borders and b/ordering play in terms of how the rationalities and practices of language resilience are put into action.

With the concept of "cross-border language," it is typically referred to minority languages spoken by one ethnic group living across two or several state borders. According to Willemyns (2002), there are two different ways to understand the connections of changing borders and languages. First, when borders shift, languages shift. This refers to historical processes where places that used to be part of the transition zone between two nations have moved into the monolingual zone on one side of the state border (cf. Frandsen 2022). The second way of understanding the connection of borders and languages is associated with language shift resulting in "erosion," meaning that "the contact situation has decisively been changed in the course of history although the 'language border' (in the traditional sense) has not changed its course" (Willemyns 2002, p. 38). As Andersen and Prokkola (2022, p. 6ff.) highlight, borderlands are ecological, political and social environments where local people have long histories of coping with and within the structures of two states. Shifting borders often leads into linguistic marginalization, and for minority languages to survive, local-level revitalization work is required.

"Language revitalization" refers to activity that has protective function for the indigenous communities and their members, conveying resilience where significant disparities otherwise exist (Fitzgerald 2017). There are several rationalities for why saving minority languages is important, such as questions concerning identity-formation, cultural diversity, educational principles and the sovereignty of minority groups (Ridanpää 2018). Along with arguments such as "cultural diversity is as necessary for humankind as biodiversity is for nature" (Skutnabb-Kangas 2002, p. 2), many scholars across disciplinary boundaries have underscored how languages in themselves are to be considered cultural treasures. When languages diminish, part of human knowledge and cultural heterogeneity is lost with it. Within this context, the idea of resilience comes close to the questions and definitions regarding sustainability (see Clark-Joseph and Joseph 2020, pp. 146–151). Correspondingly, the Canadian government, for instance, has recognized language revitalization being an issue of health and wellbeing of individuals and communities (Duff and Duanduan 2009).

Minority languages are ingredients in wider societal discussions about the confrontations between the ideology of nationalism and liberal democracy (May 2012). Language emancipation often requires linguistic standardization, which, instead of the needs of minority language speakers, serves the political interests of nation-states (see Lane 2011). Hence, from the viewpoint of nation-state ideologies the very idea of revitalizing borderland languages is problematic. On the other hand, a more predominant opinion is that language revitalization, first and foremost, can be utilized as a means by which it is possible to detach from past ideologies of colonialism (see Hermes 2012). For national minorities, language revitalization often works as a tool for rediscovering an othered identity and a sense of pride (Ridanpää 2017). In this way, the endeavors to "save the language" may gain a major symbolic value (Sallabank 2013).

Contextualizing the Meänkieli language

After the Grand Duchy of Finland had been separated from Sweden in the aftermath of the war of 1808–1809, the political movement of Fennomans started to work with intentions to raise the respect and social status of the Finnish language. Major changes in legislation, particularly the language manifesto in 1863, initiated increasing organizing and cultural-political activity in civil society that ultimately led the Grand Duchy of Finland to gain independence in 1917. At the same time, the Finnish-speaking population on the Swedish side of the Torne Valley region became a linguistic and national minority. Finland was understood to represent a threat to national security, and Finnish-speaking people living in northern Sweden were perceived as an ethnically inferior population, an internal "other" to the Swedish-speaking majority (Elenius 2002) – "a fact" that was ostensibly "proved" by the racial studies of Swedish anthropologists (Heith 2012). Through institutional control, particularly the school system, the marginalization of the Swedish Torne Valley became a concrete, everyday feature of people's lives in the area. During the first half of the 20th century, speaking Finnish at school was forbidden in northern Sweden (see Júlíusdóttir 2007, p. 41).

The hostility continued throughout the early 20th century until relations with an independent Finland improved (Hult 2004, p. 188). During the years of "Swedification," Finnish heritage became something to be ashamed of, something that needed to be hidden; for instance, exchanging Finnish surnames for Swedish ones became highly popular. At the same time, the number of Finnish speakers decreased remarkably (see Prokkola 2009, p. 28). Bilingualism was conceived as "halflingualism," an unsatisfactory proficiency in both languages and something to be ashamed of (Ahola 2006, p. 28). As the writer Mikael Niemi in his break-through novel *Popularmusik från Vittula* (*Popular Music from Vittula*), published in 2000, insightfully describes the regional history of linguistic otherness: "We spoke with a Finnish accent without being Finnish, and we spoke without a Swedish accent without being Swedish. We were nothing" (Niemi 2003, p. 49).

The "ethnic renaissance" of the late 20th century resulted in a major change in local social self-esteem in the region, and a new impetus to preserve the regional culture and language arose (see Winsa 2005). The organization Svenska Tornedalingars Riksförbund (STR-T) was established in 1981 for the revitalization of Tornedalian culture and language. In terms of rising regional awareness, the major role was played by culture activist/writer Bengt Pohjanen, who has worked in several ways towards language revitalization. As a symbolic marker for regional belonging, the language spoken was not called Finnish anymore, but "Meänkieli," literally "our language," and quickly "meän," "our," evolved into a keyword around which all new cultural and social activity became entwined. Similar kind of ethnic revivals, as well as civil rights movements, took place all around the world during the late 20th century (Fishman 1985).

According to *Ethnologue: Languages of the World* (2018), there are approximately 30,000 Meänkieli speakers on the Swedish side of Torne River, but the exact number of speakers is highly difficult to estimate, as people are often uncertain whether the language they use can actually be defined Meänkieli or not. In addition, opinions vary on whether the dialect spoken on the Finnish side of Torne Valley should be also called Meänkieli or not. In Finland Meänkieli is often defined as a dialect of Finnish, characterized by the extensive use of h sounds, the loaning of Swedish words, and a certain form of regression compared with Finnish spoken in today's Finland (see Vaattovaara 2009). In addition, switching languages from Meänkieli/Finnish to Swedish and back again even during a single sentence is typical for Meänkieli speakers and can be considered a specific characteristic of the language. Although the distinct status of Meänkieli has often been questioned (Piasecki 2014, p. 13) and in Finland Meänkieli is defined as a dialect of Finnish, in 2000 Meänkieli was granted official status as a minority language in five municipalities in northern Sweden: Gällivare, Kiruna, Haparanda, Pajala and Övertorneå.

According to Prokkola (2009), the borderland identity of Torne Valley cannot be explained by any single attribute such as language or ethnicity. However, minority language work has been a symbolic cornerstone on which the cultural work striving for regional self-esteem has been based. After Mikael Niemi's novel *Popularmusik från Vittula*, mentioned earlier, became a bestseller, the history of the Meänkieli-speaking minority was acknowledged for the first time more widely in both Sweden and Finland. Although it was a coincidence that the book was published the same year when the Meänkieli language act came into force, for the language activists this coincidence was extremely important, no matter how negative the image of the region in this dark ironic novel was:

> As a citizen of Pajala, you were inferior – that was clear from the very beginning. Skåne, in the far south, came first in the atlas, printed on an extra-large scale, completely covered in red lines denoting main roads and black dots representing towns and villages. Then came the other provinces on a normal scale, moving farther north page by page. Last of all was Northern Norrland, on an extra-small scale in order to fit onto the page, but even so there were hardly any dots at all. Almost at the very top of the map was Pajala, surrounded by brown-colored tundra, and that was where we lived. If you turned back to the front you could see that the Skåne was in fact the same size as Northern Norrland, but colored green by all that confoundedly fertile farming land. It was many years before the penny dropped and I realized that Skåne, the whole of our most southerly province, would fit comfortably between Haparanda and Boden.
>
> (Niemi 2003, pp. 46–47)

At the same time, the success of the novel launched a new interpretative layer for how the theme of Meänkieli as a minority language and its connection to regional identity were to be approached (see more in Ridanpää 2019).

As Andersen and Prokkola (2022, p. 8) bring forward, the cultural resilience of minority groups is inherently entangled with the question of how the groups manage to survive as a distinct cultural community, thus implying how the identity-formation, a certain form of "construction of distinctiveness," works for the benefit of language resilience. From the viewpoint of regional geography, one interesting reproduction of cultural heritage has been the concept of Meänmaa, "our land," launched by culture activist and writer Bengt Pohjanen. Along the increasing cross-border activism, the concept of Meänmaa, originally referring to the Swedish side of Torne Valley, started to refer to regions on both sides of the river. In prevailing cartographic illustrations, territorial shape for the map of Meänmaa borderland is taken from the map drawn by Finnish linguist Martti Airila back in 1912 in his historical study of the dialect of Torne (Prokkola and Ridanpää 2011, pp. 781–782). Meänmaa has not been recognized on any governmental level, but it is nevertheless possible to get granted a (symbolic) passport of Meänmaa, whose sole practical purpose is that locals can get a discount from local shops. According to Heith (2018, p. 105), Pohjanen has used naming, mapping and symbolic elements connoting Tornedalian culture to decolonize Meänmaa.

Although several local culture activists consider Pohjanen's borderland brand of Meänmaa artificial and unnecessary, and rather prefer using the traditional names of Torne Valley or Torne River Valley (Ridanpää 2018, p. 194), by scripting local stories and documenting folklore Pohjanen has created an archive of borderland heritage, which provides a resource for regional culture industries (Prokkola and Ridanpää 2011, p. 782). Pohjanen's work is a good of people's resilience, a social impact that starts from the grassroots, from bottom up all the way to the higher institutional levels (see Andersen and Prokkola 2022, p. 5). Pohjanen's writings on the borderland culture and heritage have provided material that has been commercialized in EU-funded cross-border projects, such as INTERREG III A North, at the same time offering both resources and symbolic legitimation for language revitalization work. The implementation of the School of Language and Culture, established alongside the elementary school in Pello during the program period of 2000–2006, has been one of the most long-standing cross-border cooperation initiatives conducted with the help of INTERREG funding. From the local point of view, the establishment of a cross-border language school was important, since at the same time the maintenance of basic communal infrastructures and services was enabled (Prokkola et al. 2015, p. 111). One of the central goals of the school initiative was to raise regional consciousness and identity, teaching pupils alternative regional histographies and cultures.

Methodology for hearing stories

As Andersen and Prokkola (2022, p. 6ff.) underscore, the study of borderland resilience includes questions concerning the experiences and narratives of borders, social relations and belongings. People make sense of spatial belonging, of who they are, what kinds of communities they belong to, through telling stories based on their experiences; stories may also be used to narrativize their position in relation to others, that is, for assembling people's spatial bearings and identities into a narrative with a plot (see Rycroft and Jenness 2012; Rose 2016). Along with its instrumental and communicative roles, a key function of language is to maintain group identity and shared feelings of belonging, and thus language has a key role in the process of how spatial identities are constructed (Edwards 2010), while identity narratives function as performances of spatial belonging.

Here it is essential to separate the concepts of "identity narrative" and "narrative identity." While the term "identity narrative" refers to the stories of peoples' self-conceptions (Yngvesson and Mahoney 2000), the term "narrative identity" is commonly used to refer to the ways that individuals construct their personal stories in certain social circumstances, as well as to the ways in which communities construct their spatial identities through stories (Somers 1994). In the case of language revitalization, the essential question concerns how it is enabled that the voices and stories of minority groups become heard, the identity narratives rooted in the heritage of marginalization and othering, today being characterized by simultaneous acceptance of uncertainty and hope for the better future. In this research, "hearing stories" is implemented using a group discussion method. In minority studies, group discussions have been considered a unique methodological route through which oppressed minorities are offered a possibility to make their voices heard (Booth and Booth 1996), a method through which the multiplicity of shared as well as contested narratives can be discovered (Price 2010).

The following analysis bases on eight group discussions, conducted on the Swedish side of the Torne Valley between September 2015 and February 2016. The participants were people who are (inter-)actively working – some directly, some less directly – with the revitalization of Meänkieli, such as members of village associations and cultural associations, teachers from different levels of education, local radio reporters, choir members and musicians. Methodologically speaking, the groups and interviewed persons were not treated as key informants, but rather as community members who have formal or informal personal experiences from language revitalization work. The groups and interviewed persons are unnamed for confidentiality. For each discussion, a separate thematic framework was prepared, containing questions relating to the specific activities of the interviewed group (music, radio broadcasting, teaching, religion and so on). In addition, there were some highly general questions that were directed to all interviewed groups concerning the local languages and identity. The key questions of this kind were: (1) How do you see the work on Meänkieli language revitalization, and (2) how does the issue of minority language come up or need to be acknowledged in your work? All discussions were conducted in Meänkieli, although there were natural alterations in language use depending on the backgrounds of different discussants.

Revitalizing the Meänkieli language

How to revitalize minority languages?

During the group discussions, it quickly became clear that there were two approaches to the most crucial question in language revitalization – that is, how to do it? Some discussants started their argumentation from the premise of how necessary it is to institutionalize the language, while others considered the most relevant issue is an increasing positive attitude towards the language among the local people. As Meänkieli has been a colloquial language with no institutional status, used only in daily conversations, seeing Meänkieli in printed form has a major symbolic value in terms of how the social status of Meänkieli becomes legitimized:

> Woman 1: Nobody demands anything in Meänkieli because Meänkieli speakers cannot necessarily read Meänkieli. All the Meänkieli speakers can read Swedish, so there hasn't been any need for it. Only so, as we have spoken, that in symbolic manner it is nice that we can see Meänkieli, like for example in municipality webpages there can be something in Meänkieli. So, it's important that it's visible, but there has never been a situation in which someone insisted that things must be written also in Meänkieli.
>
> (Group 6)

With a history of being treated as an uncivilized people with an uncivilized language, spoken at the "wrong" side of the border, the (symbolic) institutionalization of Meänkieli is considered highly important. In similar fashion, having church services in Meänkieli was considered not only homely and cozy, but also important as such. The other argument underscores how Meänkieli, both as a skill and a part of identity, should be made attractive to the local people. It is relatively common that minority languages are, especially among youth, considered a skill to be ashamed of (McCarty et al. 2009, pp. 300–302). As is widely recognized, this is also one of the key reasons why several minority languages are vanishing in the first place. However, it is axiomatic that revitalizing endangered languages require having and teaching a positive attitude no matter how demanding or desperate the work itself feels. A Meänkieli language teacher commented:

> Woman 1: When you're working with Meänkieli it is important that you give a positive image of it, like it's fun and whatever you do with Meänkieli, should somehow be fun so that you get a feeling like: "Wow, I wanna learn that too." A kind of interest and fascination rises, which helps the language survive and children are seriously willing to learn it and try to find those friends whom with it is possible to speak the language. So, for the sake of revitalization it is extremely important that it is made somehow interesting. And whatever you do with Meänkieli, it must be fun.
>
> (Group 6)

Keeping language teaching entertaining is in connection with another interesting question about how humor has been used in the revitalization of Meänkieli, for example when Bengt Pohjanen's grammar education was broadcasted in local radio (see Ridanpää 2018, p. 195). On the other hand, the interconnections between humor, identity, minority languages and also otherness often relate to the question of whether the language can be comical in itself. For example, speaking English with a Scottish accent has in some contexts been used as a rhetorical vehicle to provoke laughter, by using the dialect as if it were humorous in itself. Does Meänkieli sound "funny" in a way that it is easy to mock or whether listeners may hear some heart-warming tone in it? According to discussants, both reactions have been noticed (see more in Ridanpää 2017). This illustrates how emotionally loaded topics Meänkieli language and revitalization are. In the 1960s, basically all the children on the Swedish side of Torne Valley spoke Finnish/Meänkieli, but after that the decline in the number of Meänkieli-speaking children declined dramatically:

> *Man 2:*   In regrettable manner all those who speak and have roots here say that "dear me, for why we did not teach our children."
> *Man 1:*   Yes, that's the way how it is and so say the children too, blame their parents.
> *Man 3:*   Yes, I've heard hundreds and hundreds who say so.
>
> (Group 3)

The case of Meänkieli illustrates how "we-feelings," as spatially embedded emotions, are not only about the feelings of belonging, but also about the feelings of being different (see Richter 2015). One generation skipped learning the language because their parents were ashamed of their historical background. In trans-national families, different languages are used for different purposes (Soler and Zabrodskaja 2017), but in case of Finnish/Meänkieli, avoiding speaking the language to your children was axiomatic, an in-built norm: speaking Finnish was not an option.

Rivalry of belonging

The geography of "linguistic differentiation," a concept referring to how languages are conceived as discrete, bounded entities, embedded in the politics of a region and its observers (Irvine and Gal 2000), plays an essential role in case of Meänkieli. Meänkieli is considered a language spoken only in the Torne Valley, while the Torne Valley is geographically defined as a river basin of the Torne river. This has caused conflicting opinions about who should be entitled as genuine Meänkieli speakers, and who should not. As one woman exemplified: "These people from ore-mining fields, they think that they are from the Torne Valley." According to topographical definitions, the ore-mining fields of northern Sweden are not a part of the Torne Valley. The headwaters of a border river between Sweden and Finland do not come from ore-mining fields, so despite the fact that the historical background of the people living there is connected to Finland, in terms of language politics, they are considered "outsiders." In this "rivalry of belonging," Meänkieli becomes in a symbolic way defined as "a border river language," an argument that some people find essential to keep up, while other discussants, in contrast, considered a substantially annoying feature of Meänkieli language preservation work:

> Woman 1: What has bothered me a bit is that people are terribly restricted. Like if you belong to this group, you cannot belong to that group, because you are for example from Finland.
> . . .
> It shouldn't matter. The main thing is that we speak something. It's not relevant whether we speak more Finnish words or more Swedish words.
> (Group 7)

As Prokkola (2009) argues, in the case of the Torne Valley the borderland identity becomes composed of various overlapping and contradictory voices and narrations that people have. The geographical definition of the Torne Valley is inseparably attached to the definition of Meänkieli and how the connection between the language and identity is understood. However, according to Bradley (2019, p. 511), one problem in minority language resilience thinking is that people conceive that some territories were designated belonging to some specific minority groups as "natural order." For Bengt Pohjanen establishing and promoting the territorial shape of Meänmaa borderland has been a part of revitalization work, but there still are a variety of contradictory narrations over how the history of Meänkieli should be comprehended. In one group discussion, the societal status of Meänkieli was defended by re-narrating the history of the Finnish language:

> Man 1: It has been acknowledged that, ever since the Torne Valley has been populated, say back to the 10th–11th century, the language spoken here was Meänkieli. . . . Finnish, the national language of Finland, it was not before the 16th century when Mikael Agricola started mucking around with it. It was not until the middle of 19th century when Finnish became a public, national language. It is so that if you take a more in-depth look, the national language of Finland is a dialect of Meänkieli, if you may, not the other way around.
>
> (Group 8)

Promoting Meänkieli as an endangered cross-border language is also a rivalry between various aspects of understanding the history of the Torne Valley and the role of Finland and Finnish language within it. Whether the discussant in the previous example is serious or merely telling a story in jest is irrelevant here. According to discussants, Meänkieli contains something pure and original that other, officially acknowledged state languages do not have. The modern nation-state system, along with redefined state borders, is understood as breaking the "natural order" and by that means establishing the incorrect narratives of linguistic history (see also Andersen 2022). The feeling of being othered is constantly embedded and self-imposed when discussing the societal position of Meänkieli, and, interestingly, as a border language, this bitterness is simultaneously directed both at the history of Swedification as well as at the history of Finnish society.

Critique of revitalization

What came out as a surprise during the group discussions was that several people who were actively involved in language revitalization work still considered Meänkieli as "an invented language": "I have always said that I speak Finnish, but it was before that, when Meänkieli was invented." On the other hand, it is important to underscore that the many issues that were figured out along the ethnic renaissance in the 1980s were cultural inventions of a certain kind. Culture workers acknowledge that "meän," "our," has turned into a borderland brand that is used in several fields of cultural and communal activities, and according to some discussants, no matter how important they consider the value of minority language work, the identity branding has simply gone too far:

> Woman 1: Some say that it's too much, like in all occasions it is Meänkieli, Meänkieli.
>
> (Group 5)

> Woman 1: I am not sure whether I'm a right person to say, since I'm a little bit like a Finn and I speak Finnish and try to teach Finnish. Sometimes

it feels like it has turned into a kind of myth, that municipalities must have Meänkieli and it must be used.

(Group 2)

Calling "Tornedalian Finska" Meänkieli has been a strategic move towards more vivid borderland identity, and it is understandable that for some, even those being actively involved in revitalization work, the artificiality may feel irritating. As mentioned, whereas in Finland Meänkieli is understood as a dialect of Finnish, in Sweden the categorical division between Finnish and Meänkieli language is important, or to be precise, this difference is a fundamental basis of the whole minority language and identity revitalization work. In a similar way, some level of institutionalization of Meänkieli is crucial, and as underscored in this chapter, in the case of minority language work the role of key regional activists is essential. In case of the Torne Valley, Bengt Pohjanen has been a key person in the mobilization of borderland culture, language and identity work. One of the most important symbolic signposts in his language revitalization work was when he published the grammar for Meänkieli in 1996. Yet, although Pohjanen's active language work has led into the legitimation of Meänkieli's official status as a minority language, people also talk about how standardizing everyday language may actually hinder its survival:

Man 1:   When that grammar came, when they started working with that book, . . . I have sometimes pondered it like by that means it is made more difficult than it actually is. When you start phrasing like, "this must be in that form, this is correct, this is wrong," then you prevent the language from developing, I think.

Woman 1:   Yes, then the interest stops.
. . .
Man 1:   I believe that if language must survive, it must be. . .
Man 2:   Free.
Man 1:   Yes, free. And it isn't that serious if something is said wrong.

(Group 6)

It is presumable that the speaker in the previous example does not think that "let it be" is the best method of saving the language, but what it illustrates is that opinions about how seriously the language work should be taken or forced into action vary. Meänkieli is a cornerstone of the Torne Valley's borderland identity, but although the revitalization of Meänkieli has been recognized as a matter of regional pride, the language revitalization work can also be interpreted, in contradictory manner, happening at the cost of the appreciation of Finnish identity and roots:

Man 1: Linguistically it seems like that Meänkieli, as a name, has gained a kind of monopoly here in Swedish side, on the western side of the border. Nobody speaks about Finnish. . . .

As a name, Meänkieli has then overtook . . . that what language is spoken here, at this side of the border. Haparanda is actually that island, in where we still say, that we speak Finnish, but in other municipalities up north, the title is Meänkieli. . . .

It is something that enables to deny own roots, when being Meänkieli speakers. On no account being Finnish. Not even from our roots. It is the same thing there in Finnmark, Norway. It is only talked about Kven people, . . . not anymore about Finnmark people or the Finns. They are Kvens nowadays. No matter whether they wanted it or not.

(Group 1)

Whereas Pohjanen's borderland regionalism has contested hegemonic nationalistic narratives, the discussant here emphasizes how Meänkieli, as an INTERREG-funded identity brand, has turned into a hegemonic discourse at the local level, and the narrations of "who we are" (sense of pride) and "where we come from" (sense of shame) are not in harmony. This resonates with a development what Liebich (2019) terms a "boomerang effect," how kin-state cross-border activism may divide the regional communities it seeks to unite. The language revitalization work obviously serves to benefit identity work, but as a discussant conceives here, in a slightly contradicting manner, with the cost of the decreasing appreciation of Finnish identity and roots, that is, part of the history of Meänkieli.

Conclusions

In the case of minority languages, resilience thinking refers to recognizing and accepting the irreversible development of language loss and still being able to live with it. In most cases, globalization is considered the biggest threat for the survival of minority languages, which has also been noticed in the Torne Valley. According to discussants, the language typically used in cross-border cultural cooperation is not Meänkieli, nor Swedish, nor Finnish, but English. Some people consider this a threat to minority language revitalizing, while for others this kind of conservation feels unnecessary and hypocritical. However, in the case of border languages the main challenges still come from the state-centric ideologies and practices. As Prokkola (2009, p. 24) formulates this,

in the case of a political boundary being drawn through a culturally and linguistically coherent region, as with the Finnish-Swedish border, the border region becomes a transition zone where state institutions and the practices of spatial socialization collide with the local cultures.

As the scholars in minority language studies argue, language revitalization is valuable in itself, but in case of Meänkieli, language revitalization is interconnected, partly directly, partly indirectly, with several other cross-border activities, such as tourism development (see Prokkola 2007). For many people involved in language revitalization work and other cross-border cultural activities, the fundamental issue has been the symbolic acknowledgement of Meänkieli as an integral part of how borderland identity is understood. However, it is important to emphasize that the histories of Meänkieli and Meänmaa borderland are identity narratives based purely on regionalist activism and not shared by all the inhabitants living across the border, nor even by the activists involved in revitalization work. As shown here, there are multiple understandings about the nature of local heritage and recognizable disharmonies between different conceptions of how and why Meänkieli should be revitalized. One important aspect to these questions is to acknowledge that while recent "fast stress events" have disrupted social and cultural activities in borderlands in multiple ways (Andersen and Prokkola 2022, p. 1), revitalizing minority languages is one of the best examples of what resilience thinking means in practice and how cultural resilience is enforced into action.

References

Ahola, L., 2006. 'Asenteellista menoa – tornionlaaksolaiset nuoret ja vähemmistökieli' ('Prejudiced going – Tornedalian youth and minority language'). *In:* O. Haurinen and H. Sulkala, eds. *Tutkielmia vähemmistökielistä Jäämereltä Liivinrantaan: Vähemmistökielten tutkimus- ja koulutusverkoston raportti V* (Researches from the arctic ocean to liivinranta: report V of minority language research and education network). Oulu: Oulu University Press, 15–29.

Andersen, D.J., 2022. Line-practice as resilience strategy: the Istrian experience. *In:* D.J. Andersen and E-K. Prokkola, eds. *Borderlands resilience: transitions, adaptation, and resistance at borders.* London: Routledge, 166–181.

Andersen, D.J. and Prokkola, E-K., 2022. Introduction: embedding borderlands resilience. *In:* D.J. Andersen and E-K. Prokkola, eds. *Borderlands resilience – transitions, adaption and resistance at borders.* London and New York: Routledge, 1–18.

Booth, T. and Booth, W., 1996. Sounds of silence: narrative research with inarticulate subjects. *Disability & Society*, 11 (1), 55–70.

Bradley, D., 2019. Resilience for minority languages. *In: The Palgrave handbook of minority languages and communities.* London: Palgrave Macmillan, 509–530.

Clark-Joseph, A.D. and Joseph, B.D., 2020. The economics of language diversity and language resilience in the Balkans. *In:* C.B. Vigouroux and S.S. Mufwene, eds. *Bridging linguistics and economics.* Cambridge: Cambridge University Press, 140–157.

Dorian, N.C., 1994. Purism vs. compromise in language revitalization and language revival. *Language in Society*, 23 (4), 479–494.

Duff, P.A. and Duanduan, L., 2009. Indigenous, minority, and heritage language education in Canada: policies, contexts, and issues. *Canadian Modern Language Review*, 66 (1), 1–8.

Edwards, J., 2010. *Minority languages and group identity.* Amsterdam: John Benjamins BV.

Elenius, L., 2002. A place in the memory of nation. Minority policy towards the finnish speakers in Sweden and Norway. *Acta Borealia*, 19, 103–123.

Fishman, J.A., 1985. *The rise and fall of the ethnic revival.* Berlin: Mouton de Gruyter.

Fishman, J.A., 1991. *Reversing language shift: theoretical and empirical foundations of assistance to threatened languages.* Clevedon: Multilingual Matters.

Fitzgerald, C.M., 2017. Understanding language vitality and reclamation as resilience: a framework for language endangerment and 'loss' (commentary on Mufwene). *Language*, 93 (4), e280–e297.

Frandsen, S.B., 2022. Schleswig. From a land-in-between to a national borderland. *In:* D.J. Andersen and E-K. Prokkola, eds. *Borderlands resilience: transitions, adaptation, and resistance at borders.* London: Routledge, 121–136.

Heith, A., 2012. Ethnicity, cultural identity and bordering: a tornedalian negro. *Folklore*, 52, 85–108.

Heith, A., 2018. Ethnofuturism and place-making: Bengt Pohjanen's construction of Meänmaa. *Journal of Northern Studies*, 12 (1), 93–109.

Hermes, M., 2012. Indigenous language revitalization and documentation in the United States: collaboration despite colonialism. *Language and Linguistics Compass*, 6 (3), 131–142.

Hult, F.M., 2004. Planning for multilingualism and minority language rights in Sweden. *Language Policy*, 3 (2), 181–201.

Irvine, J.T. and Gal, S., 2000. Language ideology and linguistic differentiation. *In:* P.V. Kroskrity, ed., *Regimes of language.* Santa Fe, CA: School of American Research Press, 35–83.

Júlíusdóttir, M., 2007. Culture, cultural economy and gender in processes of place reinvention. *In:* T. Nyseth and B. Granås, eds. *Place reinvention in the North – dynamics and governance perspectives.* Stockholm: Nordregio, 39–53.

Lane, P., 2011. The birth of the Kven language in Norway: emancipation through state recognition. *International Journal of the Sociology of Language*, 209, 57–74.

Liebich, A., 2019. The "boomerang effect" of kin-state activism: cross-border ties and the securitization of kin minorities. *Journal of Borderlands Studies*, 34 (5), 665–684.

May, S., 2012. *Language and minority rights: ethnicity, nationalism and the politics of language.* New York: Routledge.

McCarty, T.L., Romero-Little, M.E., Warhol, L., and Zepeda, O., 2009. Indigenous youth as language policy makers. *Journal of Language, Identity, and Education*, 8 (5), 291–306.

Niemi, M., 2003. *Popular music from vittula* (trans. L. Thompson). New York: Seven Stories Press.

Piasecki, K., 2014. The birth of new ethnoses: examples from Northern Europe. Our Europe. *Ethnography – Ethnology – Anthropology of Culture*, 3, 7–20.

Price, P.L., 2010. Cultural geography and the stories we tell ourselves. *Cultural Geographies*, 17 (2), 203–210.

Prokkola, E-K., 2007. Cross-border regionalization and tourism development at the Swedish-Finnish border: "destination arctic circle". *Scandinavian Journal of Hospitality and Tourism*, 7 (2), 120–138.

Prokkola, E-K., 2009. Unfixing borderland identity: border performances and narratives in the construction of self. *Journal of Borderlands Studies*, 24 (3), 21–38.

Prokkola, E-K. and Ridanpää, J., 2011. Following the plot of Bengt Pohjanen's Meänmaa: narrativization as a process of creating regional identity. *Social & Cultural Geography*, 12 (7), 775–791.

Prokkola, E-K., Zimmerbauer, K., and Jakola, F., 2015. Performance of regional identity in the implementation of European cross-border initiatives. *European Urban and Regional Studies*, 22 (1), 104–117.

Richter, M., 2015. Can you feel the difference? emotions as an analytical lens. *Geographica Helvetica*, 70 (2), 141–148.

Ridanpää, J., 2017. Narrativizing (and laughing) spatial identities together in Meänkieli-speaking minorities. *Geoforum*, 83, 60–70.

Ridanpää, J., 2018. Why save a minority language? Meänkieli and rationales of language revitalization. *Fennia-International Journal of Geography*, 196 (2), 187–203.

Ridanpää, J., 2019. Dark humor, irony, and the collaborative narrativizations of regional belonging. *GeoHumanities*, 5 (1), 69–85.

Rose, M., 2016. A place for other stories: authorship and evidence in experimental times. *GeoHumanities*, 2 (1), 132–148.

Rycroft, S. and Jenness, R., 2012. JB priestley: bradford and a provincial narrative of England, 1913–1933. *Social & Cultural Geography*, 13 (8), 957–976.

Sallabank, J., 2013. *Attitudes to endangered languages: identities and policies.* Cambridge: Cambridge University Press.

Skutnabb-Kangas, T., 2002. *Language policies and education: the role of education in destroying or supporting the world's linguistic diversity.* Plenary paper presented at World Congress on Language Policies, Barcelona, Spain, 16–20 April.

Soler, J. and Zabrodskaja, A., 2017. New spaces of new speaker profiles: exploring language ideologies in transnational multilingual families. *Language in Society*, 46 (4), 547–566.

Somers, M.R., 1994. The narrative constitution of identity: a relational and network approach. *Theory and Society*, 23, 605–649.

Vaattovaara, J., 2009. *Meän tapa puhua: Tornionlaakso pellolaisnuorten subjektiivisena paikkana ja murrealueena* (Our way of speaking: Torne Valley as a subjective place and dialect region for the youth in Pello). Helsinki: Suomalaisen Kirjallisuuden Seura.

Valentine, G. and Skelton, T., 2007. The right to be heard: citizenship and language. *Political Geography*, 26 (2), 121–140.

Willemyns, R., 2002. The dutch-french language border in Belgium. *Journal of Multilingual and Multicultural Development*, 23 (1–2), 36–49.

Winsa, B., 2005. Language policies: instruments in cultural development and well-being. *International Journal of Circumpolar Health*, 64 (2), 170–183.

Yngvesson, B. and Mahoney, M.A., 2000. 'As one should, ought and wants to be' belonging and authenticity in identity narratives. *Theory, Culture & Society*, 17 (6), 77–110.

10 A resilient *Bel Paese?* Investigating an Italian diasporic translocality between France and Luxembourg

Christian Lamour and Paul Blanchemanche[1]

Introduction

Resilience is "the ability of groups or communities to cope with external stresses and disturbances as a result of social, political and environmental change" (Adger 2000, p. 347). It can be defined both as a response to the challenges faced by a system and as a property of this system (Lang 2010). This response and this property can express both a resistance and a willingness of people to imagine ways of living that differ to the ones imposed by external forces (Andersen and Prokkola 2022; Chandler and Reid 2016). Urban and regional resilience can be conceived in different spatial contexts, including cross-border regions (Prokkola 2019). It can include the presence of cultural parameters enabling challenges to be overcome in a creative way (Bristow 2013; Hudson 2001). One way to approach the cultural dimension of spatial resilience across state borders is to pay attention to the evolving use of space by diasporas. These communities have been able to reprocess a sense of identity across time and space by linking their region or country of origin and their places of everyday life, despite the changing socio-economic conditions in these places. An investigation of diasporas can be key to grasping the resilience of "translocalities" (Appadurai 1996a, p. 42) at the cross-border scale; that is, the adaptability of material and imaginary geographies involving producers and consumers of transnational culture sharing a common space across borders.

This chapter explores how a film festival – which is an annual ritual event – can be instrumental in the resilience of a translocality within a cross-border regional environment and in the course of the metropolitan process over the past 40 years. The basis here is an examination of the Italian Villerupt film festival, which takes place in the border region connecting France and Luxembourg where a community of Italian descendants has been living for a century. Literature on spatial resilience, translocalities and diasporic space is first discussed, then the arguments, case study and methodology are presented. The results are approached in three parts. First, there is a special focus on the meaning of the traditional cross-border Italian translocality during the long 20th century. Second, the launch of the Italian film festival of Villerupt is approached as a response by an Italian translocality

DOI: 10.4324/9781003131328-13

resilient to the multiple challenges threatening its existence. Third, the sustainability of the festival is conceived as the ability of the Italian translocal community to gather a series of new resources beyond state borders, with a view to securing its sustainability. The Villerupt festival is presented as a response to the challenges facing the diasporic space and as a property of this diasporic space to sustain itself in a time of change. The changes have not only affected the economic resources of this diaspora, but also an ensemble of cultural parameters beyond state borders and at different scales.

The resilience of urban transnational spaces: Exploring translocalities beyond neoliberal alignments and across state borders

The resilience of communities in space including diasporas shows the interrelations existing between social systems and the resources found in their environment across borders (Adger 2000; Auty 1998). The concept of spatial resilience has been criticized for three aspects: First, its conservative, normative and top-down use by policymakers in relation to neoliberalism, which requires an adaptability of social groups in competition during the latest phase of capitalist accumulation. Second, a willingness to separate on one hand the localized adaptability of communities such as diasporas to overcome problems, and on the other, the multi-scalar regulation context, where crises and structural changes are born, developed and circulated. Third, the presence of a governance network in charge of planning the best policies to secure the existence of resilient and consequently competitive regions that is an economic-driven and up-down vision of human adaptability in space (Bristow 2010; Lamour 2020; MacKinnon and Derickson 2012; Peck and Tickell 1994). One way to overcome the managerial and neoliberal trap of spatial resilience is to pay attention to the cultural identity of social groups in a multi-scalar regional environment where a complex set of resources are evolving and where resistance processes to imposed changes may occur (Andersen and Prokkola 2022; Chandler and Reid 2016; Hudson 2001; Lamour 2017, 2019a, 2019b; Lamour and Lorentz 2019, 2021).

Focusing on the evolution of a diaspora in space is a way to approach the resilience of "transnational spaces" (Jackson et al. 2004) such as borderland regions, which can be defined as the set of material and imaginary geographies based on the migration of people, objects, ideas and symbols. Diasporas are at the heart of the constant renewal of a transnational space, by bringing with them cultural artifact and ways of living that are reprocessed and used by a cultural group through time-deepened interactions with their social, economic, cultural and spatial environment. The resilience of transnational spaces experienced and enriched by a diaspora is based on the redefinition of interactions around global cultural flows produced and used in specific localities. The concept of "translocality" (Appadurai 1996a, p. 42) synthetizes this process of localized production and appropriation in space involving de-bordered cultural flows. These cultural flows have been categorized into five types by Appadurai (1996b, pp. 33–34): *Ethnoscapes,* which is the landscape of people on the move across borders, *mediascapes*, the landscape of images, sounds and scripts that are produced, consumed and interpreted beyond borders, *technoscapes*, the landscape of technologies increasingly able to cut across borders, *financescapes*, the landscape of finance flows overcoming borders at an accelerating rate and, finally, *ideoscapes*, the landscape of ideas, terms and images associated with political ideologies and counter-ideologies circulating across borders.

Italians have been associated with the production of diasporic communities in space and in the definition of translocalities since the second part of the 19th century. Some 26 million Italians left the *Bel Paese* of Italy between 1870 and 1970 (Gabaccia 2006), to which must be added the current generation of young qualified Italians leaving the country. This migration has been generally based on the search for a better socio-economic future abroad. In the past, these Italian communities were often based on localized regions of origin (Sicily, Abbruze, Marche etc.) and settled in specific urban districts and streets. In these "new" places, they could recompose their society while mixing between one generation and the next and with other cultural groups through work, education, extra-professional activities and inter-community marriages at a larger urban scale. The transformation of these Italian diasporic communities in their places of settlement has prompted many studies (Harney 2006; Fleury and Walter 2005; Fortier 2000; Gabaccia 2006; Galloro 2015). From the perspective of the resilience of de-bordered transnational spaces – and more precisely the resilience of translocalities characterized by the presence of an Italian diaspora – one must consider the permanencies and transformations of the use of space within these Italian-associated districts. As suggested by Harney (2006), three types of practices in space have characterized members of the Italian diaspora. First, everyday life routines such as the ritual evening walk, the *passegiatta*. Second, random sacred and secular rituals bringing the community together, for example religious processions or collectively watching major football matches. Third, the presence of monuments and institutions symbolizing the presence of the community in space.

The spaces of the Italian diaspora have sometimes been reimagined to "rebrand" and gentrify the neoliberal city, such as in Toronto (Harney 2006). It could thus be imagined that the resilience of Italian translocalities is simply a way to commodify past cultural identities in space in order to secure the attractiveness of cities in competition from an economic perspective. However, some institutions associated with the Italian diaspora can be key in regenerating diasporic spaces from an identity and creative perspective. This can apply to cinema, capturing images, words and scripts circulated in a global mediascape (Appadurai 1996b, p. 35). The film industry is one of the most central sectors of activity that come to mind when thinking about a global Italian mediascape in the long term (Bertellini 2010; Woodfin 2011). Italian film festivals have been created, and especially in localities where the Italian diaspora is found. These festivals can be polycentric, as is the case for instance in the US and in Australia. They also imply the existence of governance crossing state borders to obtain the resources necessary to implement them. What can an investigation of long-duration Italian film festivals tell us about the resilience of Italian translocalities across state borders, especially when they take place in some borderlands?

It is argued that Italian film festivals can reveal the resilience of everyday Italian practices, rituals and institutions in a changing multi-faceted spatial environment (Harney 2006). These festivals can be both the "response" of an Italian translocality – and a "property" of this translocality (Lang 2010) to face challenges and to capture evolving resources associated with ethnoscapes, mediascapes, ideoscapes, technoscapes and financescapes across state borders (Appadurai 1996b). The neoliberal economy and its imperatives are not external to the resilience of these diasporic practices in space, but the economy is expected to be only one element of a broader environmental context implying an evolving cultural identification with translocalities (Hudson 2001).

The research is based on the analysis of one case study: the Italian film festival of Villerupt, which takes place in the French border town of the same name located near other towns in Luxembourg, such as Esch-sur-Alzette and Dudelange (Figure 10.1). This festival started in the mid-1970s and is one of the two most important festivals dedicated to Italian cinema on French territory. It attracts up to 40,000 people annually, with screenings taking place in this part of France and in nearby urban localities in the Grand Duchy of Luxembourg. The festival is organized in a cross-border environment where an Italian diaspora has settled in different phases over the last century.

The analysis is based on interviews conducted with different stakeholders from the local Italian diaspora found in the French and Luxembourg cross-border region. One stakeholder has coordinated the Italian film festival since the beginning. Others are politicians in charge of executive functions on the two sides of the border at different spatial scales (local, regional, national and European); therefore, people whose discourses and actions are instrumental in the production of the local and regional space. All the conversations and similar used in the current chapter have been anonymized. The participants were interviewed with an open-ended questionnaire around three topics: (1) the meaning of and changes in the Italian translocality in which they were brought up, (2) the role of the Italian film festival as a way to make these Italian translocalities resilient, and (3) the function of the film festival as a property of the Italian translocality to capture changing flows of resources circulating in different "scapes" across borders.

A resilient Italian translocality beyond borders: The instrumental festival between France and Luxembourg

A traditional Italian translocality, gone with the wind. . .

When asked about their life in the Italian diasporic space between France and Luxembourg, all the interviewees responded using the past tense, even though

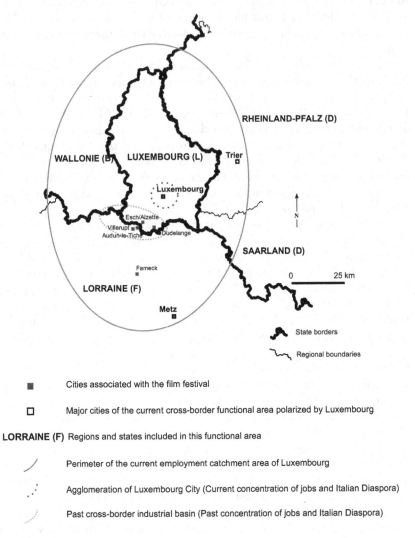

RHEINLAND-PFALZ (D)

WALLONIE (B) LUXEMBOURG (L) Trier

Luxembourg

Esch/Alzette
Villerupt
Audun-le-Tiche Dudelange

N

SAARLAND (D)

Fameck

0   25 km

LORRAINE (F)

Metz

State borders

Regional boundaries

■   Cities associated with the film festival

□   Major cities of the current cross-border functional area polarized by Luxembourg

LORRAINE (F) Regions and states included in this functional area

／ Perimeter of the current employment catchment area of Luxembourg

∴ Agglomeration of Luxembourg City (Current concentration of jobs and Italian Diaspora)

⸽ Past cross-border industrial basin (Past concentration of jobs and Italian Diaspora)

Figure 10.1 The cross-border regional space of the Italian festival of Villerupt.

the youngest was only in their early 40s. It is part of the past, and they look at it with a certain nostalgia. This past and memorized cross-border Italian translo-cality was determined in specific places where the bounds of the Italian diaspora were constantly renewed, in the same way as in other urban regions peopled by this community (Harney 2006). First, there is an ethnic homogenous residen-tial area at the local scale (e.g. the "Quartier Italie" in Dudelange or the "Rue Haute" in Herserange) within which houses regroup extended families. These families often maintain ties with the homeland through holidays, the daily use of Italian dialects or a language mix using Italian, specific culinary regional traditions and games (cards), and a keen interest in clothing fashion shared by both genders. Second, the production of this translocality also includes routinized places of daily or weekly encounters where Italian men (Italian bars) or families as a whole (Italian mass) can reinforce their sense of belonging. The cross-border dimension of this translocality is organized on the basis of two types of mobility in space. First, the mobility across the border to meet family, friends and relatives of friends belonging to the Italian diaspora, a mobility that helps to bound different residential places between the two nation-states. Second, the daily mobility towards the locations justifying the presence of the diaspora in the area: mines and steel factories exploiting localized natural resources crossing state borders; members of the Italian diaspora commuting in both directions depending on the jobs on offer and sometimes speaking Italian in the workplace. As proved in other case studies where their resilience has been researched (Adger 2000; Auty 1998), the Italian diasporic space across the France-Luxembourg border has been grounded on natural resources that have led to the local polarization of other flows circulating in distinctive "scapes" (Appadurai 1996a).

Hence, there is the ethnoscape described earlier, which includes the flow of Italians crossing borders at various scales during the different phases of the industrial process and securing the cultural dimension of their diasporic translocalities. Additionally, there is the financescape experienced by the Italian diaspora, who receive a proportion of the internationalized capital flows in relation to their professional occupation in the localized working class, whose goods are exported at the world scale. This limited capital flow means that people are concentrated in the less affluent districts near factories, where they use terraces made for them by their employers or constructed by the men of the family with resources (land and materials) provided by their employers. This Italian diaspora – with people who have experienced difficult conditions – has thereby been receptive to flows of an international ideoscape: progressive politics to improve the conditions of the working class. All the political interviewees belong to left-wing parties (ex-communist but still radical-left activists, or labor party members) with an ideological background generally determined in the Italian diasporic residential and professional space, where they experienced the difficult socio-economic conditions of their families. The interviewees' view of their Italian diasporic space is organized around parallel feelings equivalent across the border: nostalgia about friendly social relations and a festive ambiance within the Italian community and the multi-generational family; a remembrance of the hard social conditions of the Italian working class, involving physical pain and the death of family members in the mines or factories; and a sense of social injustice in terms of space, with the concentration of the Italian diasporas in specific districts. The interviewees belonging to the political class in charge of public bodies and governance networks do not look at the diasporic space as an environment peopled by a resilient community having to adapt to the current phase of capitalism, as it is often conceived as in political governance linked to resilient processes (MacKinnon and Derickson 2012). This diasporic space can be more crucible for the resilient contestation of the capitalist order as indicated by the following quote from a center-left politician. Remembrance is constitutive of the current resilience of social values in the changing economic context of the Italian diasporic translocality:

> My parents were born in Luxembourg. I am from the third generation of Italian origin, from Abruzzo, the provinces of Aquila and Teramo. My maternal grandfather was a miner. . . . My father was a factory worker and his father a factory worker. I came from the world of manual workers, and I went through very difficult times. . . . Because of their daily life, because of their work, as a child I had the experience of seeing hard work, modest pay, a modest social situation, the social differences that we saw in the city: there were workers, employees, self-employed. . . . There was still a certain difference, marked by my father's willingness as a trade unionist to commit to social justice, to solidarity. These values of solidarity and social justice were placed in my cradle and have always remained the driving force behind my social and political activity. I have always been strongly committed to civil society, so I am not limited to politics.

. . . but not lost in translation. The early festival as a revival of the working class diasporic space

The interviews with people from the France-Luxembourg border region show that the Italian translocality in which they were brought up no longer exists. However, this does not mean that the Italian diasporic space has disappeared from the cross-border region. The launch of the Italian film festival in 1976 was the reaction of a community whose cultural world was on the decline. It helped to reformulate locally two processes linked to the ethnoscape of the Italian working class diaspora: a desire for self-consciousness and emancipation within the Italian diaspora by accessing Italian artistic output that addresses meaningful issues for the working class as well as a willingness to reinvent cultural boundaries in a time of social and economic changes. This adaptation of cultural boundaries meant a transfer, from traditional gendered habits disappearing from the everyday use of space (home, homogenous diasporic districts, factories and mines) to the ritualized use of space (where the festival takes place). The festival implies that once a year, the material workforce of men is used to secure the logistics of the event and the Italian women take charge for the catering for the event with their culinary *savoir-faire*. This event attracts a localized audience as much for the sense of festivities linked to the Italian diaspora and its food as for the films that are screened, as indicated by the two following interviewees:

> When we talk about the 70s and 80s, there was a central activity: catering. The hand-made pasta of the *mamas* became a legend. People came from all over the place to eat pasta made by the *mamas* of Villerupt, and people talked more about the hand-made pasta than about the films at that time.

The spirit is always the same, it is animated. It's the party around the festival. People come, some from far away. They come to see films, they come
to eat. It's Italian food. A whole atmosphere has been created around the
festival. Not all the people come for the films; some only come for the food
and the atmosphere that surrounds the festival. . . . There is a special atmosphere, which is very popular. It's the atmosphere and the human warmth
that are most important for me. As I'm not a big cinema lover, I'm going to
see some films that show the particularities of the Italian people, the family
atmosphere that reigns in Italian families.

The reinvention of the Italian translocality by the festival – with the transfer of
practices from the everyday use of space to the ritualized use of space – is made
possible because of the coalescence of an ideoscape and an mediascape meaningful for the Italian diaspora, and more precisely the strength of the socialist ideology among dominant Italian film-makers, local councils and the Italian working
class found in the cross-border region. The festival was launched in Villerupt,
which is governed by the communist party, and soon became coordinated by a
local organization linked to the municipality (*La Maison des Jeunes et de la Culture*: MJC). At that time, the MJC (in the municipality and generally the whole of
France) was a place for creativity and cultural contestation movements considered
subversive by the French central state (Besse 2015). The film program during the
first few years of the Villerupt festival and the visibility of the event are related
through the circulation of radical-left images, words and scripts in the Italian
mediascape, and more precisely the importance of this ideology for the dominant
Italian film directors of the time, such as Ettore Scola and Luigi Comencini,
who came to Villerupt and secured publicity in Italy for this festival. The films
that make sense for the local diaspora in the region are Italian movies from the
"golden age" of the 1960s and 1970s, and social comedies where social class
and dominance issues are portrayed. The emblematic social comedy of that time,
which was shown at four festivals out of eight, is *Bread and Chocolate*, by Franco
Brusati (1974), with the actor Nino Manfredi (who also came to Villerupt) playing the role of an Italian working class migrant in Switzerland. The launch of the
Villerupt film festival was a ritualized response by a resilient working class Italian translocality, where the everyday diasporic practices in space are disappearing
while a direct link to Italy is getting narrower from one generation to the next.
The continuation of this festival for over 40 years represents a property of a transformed Italian translocality to use a series of resources in structurally changing
"scapes" at different scales, including cross-border.

The great escape from "scapes": A film festival as a property of a resilient Italian translocality

The festival of Villerupt had to deal with a series of crises in relation to the ethnoscape, ideoscape and mediascape that justified its launch. It could have collapsed in the mid-1980s after a year of not being held. However, its return and

strengthening in terms of the audience and the program show that this festival was not only a reaction of an Italian diasporic translocality, but also a property of this Italian translocality, able to capture new flows in order to sustain a festival that was in turn able to maintain a ritualized use of space related to the transalpine diaspora. Neoliberal capitalism, the strengthening of economic values in every sector of human activity and the collapse of traditional progressive ideologies form the new context faced by the Italian festival. It has meant the end of the century-long Italian working class ethnoscape in the area. Nevertheless, this festival – as a property of a resilient Italian diasporic space – has been repositioned in a broader ethnoscape located in a cross-border and polycentric metropolitan region.

In this region, two types of middle-class and upper-middle-class people can be found, who are also open to a diasporic film festival: first, a group of film lovers disconnected from the local diaspora and, second, a new generation of native Italians composed of hyper-qualified workers drained from Italy from the late 1990s onwards by the high-level service economy in the nearby agglomeration of Luxembourg City. This repositioning has meant a rescaling of the ritualized use of space by arranging the screening of Italian films in more locations, including the large peripheral cities of the region, such as the French regional capital cities of Nancy and Metz and the city of Luxembourg where the new Italian diaspora can be found, but also in regenerated cultural places of the Grand Duchy in the cross-border industrial basin peopled by the Italian working class diaspora (e.g. the National Audiovisual Center of Dudelange and the KulturFabrik in Esch-sur-Alzette).

This renewal was made possible because of the ability of the Italian festival to overcome the reduction of images, words and scripts, which used to be dominant in the ideoscape and mediascape meaningful to the Italian working class diaspora – that is, ideologies associated with the radical left. The late 1970s constituted the end of a golden age for Italian cinema with the death of neorealist film directors (e.g. Rossellini), the collapse of transalpine film production and the control of film production taken over by television networks dominated by the commercial media empire of Silvio Berlusconi (Schifano 2011). In terms of the ideoscape and mastering the cultural vanguard, the communist parties became progressively marginalized from the 1980s onwards and were generally replaced in local politics by socialist and socio-democratic parties able to make deals with the liberals and green parties. In parallel, the festival of Villerupt is managed by a dedicated institution (*Le Pole de l'image*), which separated from the amateur, multi-tasking and contesting MJC when control of the city was lost by the communist party in favor of the socialists. The scope of this new institution is to secure the sustainability of the festival by a double film program showing the presence of neoliberal economic imperatives, but also a willingness to secure its artistic legitimacy among top-level cultural institutions. The objective of this new institution is not to secure the "endangered species" of the artistic communist vanguard. There will be a continued program from one sector of Italian film production (comedy) to attract the general public, but a competition among a narrower segment of Italian film producers, with a special prize being awarded to a first/second-time film director offering more cerebral content. This double objective can be viewed as successful if we consider the importance of the public (the festival has 40,000 visitors), the sponsorship of the event by Luxembourg and French public agencies and the continued presence of film directors, actors and top-level representatives during the festival, from Italian ambassadors to French and Luxembourg ministers.

It could be imagined that the structural transformation of the Villerupt film festival over the past three and half decades is not the result of a resilient Italian working class translocality, but a mere adaptation of a local and professionalized film institution to the neoliberal supply-and-demand market rules of foreign film festivals. Nevertheless, some elements are continuously emphasized by the film festival in terms of securing its link to the century-old Italian working class translocality between France and Luxembourg. Villerupt and its barely regenerated working class spatial environment (in terms of architecture and society) is still at the heart of the festival. Much attention is also paid to an important element of the Italian working class diaspora: the presence of festive food based on Italian cooking and affordable for anyone. The Italian *mamas* and their kitchen cookery have disappeared due to their age and the loss of this tradition among Italian descendants, but there is continued attention paid to the open-to-all festive occasion based on Italian food at a low price. Furthermore, apart from the senior guests (ambassadors, film directors etc.), there is no process of segregation by social class. Anyone coming to the festival needs to stand in long queues of people, who must then watch films sometimes in basic and uncomfortable venues, in the same way as the first events targeting the local working class. It should also be noted that the cultural legitimacy of the festival does not prevent it from screening films with a social and political message against the mafia, fascism, the Italian state and its occupants; for example, a film shown about Berlusconi led on one occasion to the departure of an Italian ambassador attending the screening.

The posters for the festival – which have the function of attracting the interest of an audience from the cross-border and polycentric metropolitan region – also offer a constant reminder of the Italian working class diasporic translocality from which the festival stems. Attention is paid randomly to the chimneys of the steel factories constituting the original place where the class consciousness of the Italian diasporic workers was formed in the region. The posters also show members of the diasporic working class dressed up in their "Sundays clothes" in specific places used routinely by the diaspora, such as the street outside the local café or the modest terraces. One of the most representative posters symbolizing the attachment to the Italian working class translocality by the now-acclaimed festival is probably the one made for the 40th event in 2017. This poster displays a golden palm (a reference to the famous film festival of Cannes) given to an Italian couple whose physical features and clothing reflect their belonging to this diasporic working class. The factories justifying this localized diaspora in the long term are signified by a back and red curtain, the colors of which can reflect both cross-border night skies when metal being worked on came from the industrial furnaces, and the socialist ideals shared by this cross-border working class (Figure 10.2).

Figure 10.2 Poster of the 40th film festival of Villerupt.

Conclusion: Resilient diasporic translocalities, borders and scales

The economy and adapting to its prerogatives in a changing capitalist system are mentioned when the issue of spatial resilience within or across borders at the urban and regional scale is addressed (MacKinnon and Derickson 2012; Prokkola 2019). Spatial resilience can also involve the willingness of people to imagine alternative ways of living in a time of change, a willingness revealing that resilience is about the social and cultural responses of communities in space (Andersen and Prokkola 2022; Chandler and Reid 2016). The state border can accordingly be considered as a line of economic, social and cultural interface, potentially enabling access to key resources and securing the resilience of cities or regions. The scope of the chapter is not to examine the resilience of urban and regional places as a whole, but to pay attention to a diasporic urban space across borders and to examine its dynamics through the promotion of ritualized practices in space. This may offer a way to better approach the importance of culture, identity and creativity in the definition of spatial resilience (Bristow 2013; Hudson 2001). In the chapter, the Italian film festival of Villerupt has been presented as a response by and a property of a resilient Italian translocality. This yearly event has been based on the remobilization of the working class material and imaginary geographies from a daily use of space, such as interactions in homes, bars and factories, to a ritualized use of space with its interactions in the places of the festival, where the identity of the diaspora is reprocessed with a constant reminder of its working class roots. The resilience of the diasporic translocality across borders exemplified by this festival does not express the local obedience of one community to the top-down demand of stakeholders in charge of the neoliberal and competitive city. It results from an adaptation to structural changes affecting cultural flows circulating in global scapes (Appadurai 1996b) meaningful for the diaspora, and especially flows associated with the ethnoscape, mediascape and ideoscape. This adaptation can imply the capture of new resources at the cross-border and regional scale.

The state border is a constant line of interface where the Italian translocality can be regenerated depending on the stage of the resilience process. The small-scale urban agglomeration structured around the mining and steel industry where the Italian working class is rooted is where the response of a resilient translocality was born in the 1970s. The large metropolitan level, including affluent communities and audiences, public funding and top-level legitimizing cultural institutions (e.g. ministerial agencies) is where the property of the resilient Italian translocality based on a ritualized culture has been constantly expressed since the late 1980s. Hence, the example shows that the border – analyzed as a line of interface to secure spatial resilience in-between states – requires that attention is paid to the sequential dynamics of resilience within an expanding and changing urban space. In 2022, Esch-sur-Alzette, one of the small cities in the cross-border agglomeration where the original Italian translocality is found, will be one of the two European Capitals of Culture (ECoC) together with Kaunas in Lithuania. The territory of the ECoC Esch2022 will include nearby French cities in the ex-industrial basin, such as Villerupt. It will be interesting to investigate how the cross-border Italian translocality between France and Luxembourg will use this EU-wide cultural initiative to secure its resilience beyond borders for the next decade.

Note

1 This chapter is written within the framework of the "CECCUT" Jean Monnet Network sponsored by the Erasmus + Programme of the European Union (2018–2021). www.ceccut.eu/en/home/. Reference number: 599614-EPP-1-2018-1-LU-EPPJMO-NETWORK. The European Commission support for the production of this publication does not constitute an endorsement of the contents, which reflects the views only of the authors, and the Commission cannot be held responsible for any use, which may be made of the information contained therein.

References

Adger, W.N., 2000. Social and ecological resilience: are they related. *Progress in Human Geography*, 24 (3), 347–364.

Andersen, D.J. and Prokkola, E-K., 2022. Introduction: embedding borderlands resilience. *In:* D.J. Andersen and E-K. Prokkola, eds. *Borderlands resilience – transitions, adaption and resistance at borders.* London: Routledge, 1–18.

Appadurai, A., 1996a. Sovereignty without territoriality: notes for a postnational geography. *In:* P. Yaeger, ed. *The geography of identity.* Ann Arbor, MI: University of Michigan Press, 40–58.

Appadurai, A., 1996b. *Modernity at large.* Minneapolis, MN: University of Minnesota Press.

Auty, R.M., 1998. Social sustainability in mineral-driven development. *Journal of International Development*, 10 (4), 487–500.

Bertellini, G., 2010. The Atlantic Valentino. The "inimitable lover" as racialized and gendered Italian. *In:* L. Baldassar and D.R. Gabaccia, eds. *Intimacy and Italian migration: gender and domestic lives in a mobile world.* New York: Fordham University Press, 37–48.

Besse, L., 2015. 'L'action des Maisons des Jeunes et de la Culture. *Informations Sociales*, 190 (4), 26–35.

Bristow, G., 2010. Resilient regions: re-'place'ing regional competitiveness. *Cambridge Journal of Regions, Economy and Society*, 3 (1), 153–167.

Bristow, G., 2013. State spatiality and the governance of economic development in the UK: the changing role of the region. *Geopolitics*, 18 (2), 315–327.

Chandler, D. and Reid, J., 2016. *The neoliberal subject. Resilience, adaption, vulnerability.* London: Rowman and Littlefield.

Harney, N., 2006. Italian diasporas share the neighbourhood (in the English-speaking world). *Modern Italy*, 11 (1), 3–7.

Fleury, B. and Walter, J., 2005. 'Le festival du film italien de Villerupt: minoration nationale, majoration culturelle. *Cahiers de sociolinguistique*, 1 (10), 51–61.

Fortier, A.-M., 2000. *Migrant belongings: memory, space, identity.* Oxford: Berg.

Gabaccia, D.R., 2006. Global geography of 'little Italy': Italian neighbourhoods in comparative perspective. *Modern Italy*, 11 (1), 9–24.

Galloro, P.-D., 2015. 'Expertise de l'inclusion italienne (en)chantée ou la transformation spectaculaire d'un monstre. *Migrations Société*, 2 (12:1), 35–53.

Hudson, R., 2001. *Producing places.* New York: Guilford Press.

Jackson, P., Crang, P., and Dwyer, C., 2004. Introduction: the spaces of transnationality. *In:* P. Jackson, P. Crang and C. Dwyer, eds. *Transnational spaces.* London: Routledge, 1–23.

Lamour, C., 2017. 'The neo-Westphalian public sphere of Luxembourg: the rebordering of a mediated state democracy in a cross-border context. *Tijdschrift voor Economische en Sociale Geografie*, 108 (6), 703–717.

Lamour, C., 2019a. Popular media in the metropolitan third places: exploring the uses and gratifications of the mobile *homo œconomicus*. *International Journal of Communication*, 13, 2921–2938.

Lamour, C., 2019b. 'Researching MediaSpace in a European cross-border region: the meaning of places and the function of borders. *Communications. The European Journal of Communication Research*, 46, 253–274. DOI: 10.1515/commun-2019-2068.

Lamour, C., 2020. 'Living together at the cross-border regional scale in Europe: supra-national and trans-national identity models in the greater region. *Regional Science Policy and Practice*, 12 (5), 749–760.

Lamour, C. and Lorentz, N., 2019. 'If I were to do it all over again, should I begin with culture?' The European integration from a cultural perspective in a multi-national Grand Duchy. *Journal of Contemporary European Studies*, 27 (3), 357–374.

Lamour, C. and Lorentz, N., 2021. 'Mass media and the attraction of the arts in small-size global cities: The (re)distribution of cultural capital. *International Journal of Communication*, 15, 2335–2354.

Lang, T., 2010. 'Urban resilience and new institutional theory: a happy couple for urban and regional studies. *In:* B. Müller, ed. *German annual of spatial research and policy 2010*. Berlin: Springer, 15–24.

MacKinnon, D. and Derickson, K., 2012. From resilience to resourcefulness: a critique of resilience policy and activism. *Progress in Human Geography*, 37 (2), 253–270.

Peck, J. and Tickell, A., 1994. 'Searching for a new institutional fix: the after-Fordist crisis and the global-local disorder. *In:* A. Amin, ed. *Post-fordism: a reader*. Oxford: Blackwell, 280–315.

Prokkola, E-K., 2019. Border regional resilience in EU internal and external border areas in Finland. *European Planning Studies*, 27 (8), 1587–1606.

Schifano, L., 2011. *Le cinéma italien de 1945 à nos jours. Crise et création*. 3ème edition. Paris: Armand Colin.

Woodfin, F., 2011. *Spaesati d'Italia. Emigration in Italian national identity construction from postwar to economic miracle*. Unpublished doctoral dissertation, University of California, Berkeley, CA.

11 Line-practice as resilience strategy

The Istrian experience

Dorte Jagetic Andersen

Mi identifico con la frontiera.

—Fulvio Tomizza

Introduction

The Western Balkans is a region where the local populations' everyday life has for centuries been heavily influenced by geopolitical struggles and change. As Emilio Cocco puts it:

> In these lands, people had to learn how to deal with the everyday life implications of a semi-permanent religious, cultural and even military confrontation, in a context where the three great religious Mediterranean traditions – the Roman Catholicism, the Eastern Orthodoxy and the Sunni Islam – met and clashed (Banac 1984, p. 59).
>
> (Cocco 2010, p. 10)

Additionally, the populations in parts of the Western Balkans are still recovering from very recent violent conflict in localities, which have over the years struggled more than most for borders and maps to settle. Croatia, Serbia, Slovenia, Montenegro, New Macedonia and Bosnia-Herzegovina, the newly formed states formerly united in socialist Yugoslavia, were for centuries acting battlefields of imperial powers, and they all experienced a 20th century filled with continuous state construction, usually not of the peaceful kind (Lampe 2000).

This article sheds light on contemporary bordering practices in the everyday life of people living in the northern-most part of the Western Balkans, on the Slovenian, Croatian and Italian peninsula, Istria. As the case is with the wider region, Istria is a borderland influenced by almost constant geopolitical transition. Being part of several empires and even more different states, Istrian borders, identifications and belongings have had to rearticulate in multiple ways over the years. Nevertheless, the peninsula's inhabitants avoided becoming central to and directly involved in most conflicts between empires and nation-states in the area, and, at an everyday level, the peninsula today appears as a well-integrated regional

DOI: 10.4324/9781003131328-14

space despite its status as officially divided between three states, Croatia, Slovenia and Italy, as well as by an external Schengen border.

What is it at an everyday level that prevents conflicts from disseminating into and transforming everyday life into a battlefield of divisions in its own right – as the case has been in several other localities in the Western Balkans? In its response to the question, this chapter follows the anthology's overall thesis that resilience towards geopolitical change is related to local ways of practicing borders and identifications. Inspired by Sarah Green's (2018) notions of borders understood as "traces of lines" and "tidemarks," and by relating these notions to Cocco's (2010) emphasis on the importance of "mimicry" and "play" in the everyday practices of people on Istria, the chapter asks how people on Istria invoke borders as "lines" and, hence, how "the line" (re)appears and becomes an integral part of people's self-understanding and everyday practice. Analytically, the focus is on everyday practice of lines of division and identifications and their reiteration in local context, thus mobilizing an ethnography to capture how lines are worked with, crossed and overcome, and sometimes used strategically to articulate con-flict in otherwise integrated spaces. It thereby illuminates multiple identifications in a landscape crisscrossed by lines, and where resilience towards external stress involves the constant, active appropriation of lines.

The chapter begins by sketching out Green's notion of borders as traces of lines and tidemarks as well as the methodological considerations necessary to expose them. It then continues clarifying how these "lines" are practiced in different ways on the peninsula. The section "Questioning the ownership of lines" illus-trates how a long experience with unsettled borders opens possibilities of distanc-ing oneself from and asking questions about who owns the right to draw lines. In the section "Remembering imperial spaces," memories are exposed as ways to practice lines, and the section "Intertwining national and regional lines" refers to important identifications at play in these line-practices. Finally, "Juggling multi-ple lines" exposes us to an everyday life lived with and through a multiplicity of intersecting lines. Together these practices make up the peninsula's population's ability to, in different ways, independently upset and settle "the map" through their identifications, abilities constituting local resilience towards stress caused by geopolitical change.

Borders as traces of lines and as tidemarks in everyday life

When "tracing lines" on Istria, the chapter finds inspiration in work by bor-der scholars who approach borders as performance and practice, asking ques-tions about how borders are made to matter in the everyday life of "ordinary" people. It is almost commonplace in border studies to problematize the border less as geometrical lines drawn by state actors on maps, than as messy every-day life practice, lived by "real" people in "real" time (van Houtum et al. 2005; Rumford 2008; Parker and Vaughan-Williams 2009; Andersen and Sandberg 2012). While the role of states is recognized as central when bordering processes are studied, the argument has been that the state-centered perspective underestimates the role played by ordinary people in the borderlands. Non-state actors equally make borders by crossing them, talking about them or just relating to them in one way or another.

Hence, border research is increasingly occupied with the making of borders in people's (messy) everyday lives. However, in this landscape of everyday bordering practices we should not forget that state, imperial and other borders, each with their inherent border logics, remain important, also in everyday practice. To understand how border logics are at play in everyday life, it is helpful to take recourse to Green's article, "Lines, Traces and Tidemarks" (2018), where she problematizes borders as "lines" that appear in the form of "traces." A trace is the lack or absence of something that has already been there, a sign indicating something which is not visible, yet providing tangible, often material evidence of the existence of the thing that is absent or invisible (2018, p. 77). Taussig's example (1993) of how the blue jean carries with it traces of a colonial past because the jeans' blue color originates with the indigo plant even when the plant is no longer used to color jeans, illustrates this; the trace is a material remnant of something that once was, yet is no longer there.

Even when it is clearly reductive to confine the ontological reality of borders to that of geographical dividers between states, borders do appear in everyday life as were they lines drawn on maps by states exactly because of the absent presence of "enduring marks." Material remnants, such as police controls and custom buildings, are read by people as were they geopolitical dividing lines made on maps, thereby endowing these lines with ontological reality. The markers made on paper do work, so to speak, in people's everyday life to order an otherwise messy reality: "The act of cutting in the case of border might even be called an effort at performativity: to declare that the difference between here and not-here is a particular kind of thing (e.g. a nation, . . .)" (Green 2018, p. 75). The line understood as the trace of borders on the map is called upon not just by state actors, but also by multiple others to put things in their right place, performatively carving out distinctions used to place objects (including people) into categories of here and there, in and out, us and them, one and the other side.

Lines drawn on maps can appear as traces in material form, an insight that helps us "think about the entangled relation between symbolic, material, and legal forms" (Green 2018, p. 70). Following this line of argumentation, Green also refers to borders as "tidemarks," a metaphor for the marks left by traces in space over time. Tidemarks do not come in singular form; they are made by the motion of waves that keep returning, erasing previous tidemarks and leaving new marks in the sand. Read as tidemark, the line becomes a space imbued with subjectivity and movement, something that is reflected in areas of crossing and dwelling, a space in its continuous becoming: "the borderlands." Crang and Travlou (2001) illustrate this with reference to the palimpsest. On a piece of re-used parchment or palimpsest, what was written earlier might or might not reappear through the new writings replacing it, but it remains there to leave its mark if found. What is exposed is therefore always only fragment of marks left behind. Tidemark as metaphor thereby helps us understand how multiple lines can be at play at once, read in a timely sense, as one replacing the other, and in a spatial sense, as lines placed on top of each other and waiting to appear.

Green's notions of lines and tidemarks mobilize a distinction between reality and map, ontology and epistemology as opening for reflection, as well as criticism, play etc. on the status of the line, an opening showing how the line is always-already under construction. This way of understanding borders provides an additional argument for why we should ask questions about borderland resilience to those for whom lines matter in practice (cf. Andersen and Prokkola 2022 following Wandji 2019). Understanding borderlands resilience means grasping borders as lines and tidemarks in close-up and sustained studies of practice in whatever form it may take. To the best of our knowledge, this study is enabled only by ethnography, a method addressing Dürrschmidt's (2002) claim that "to realize how local, everyday life reproduces and negotiates existing borders while also generating new boundaries, an empirically grounded analysis of everyday life practices in border regions is crucial" (p. 124).

Aiming to expose the mundane and messy line-practices in everyday life of people living in Istria, as well as grasping related processes of coping with divisions both generating and generated by borders, I therefore set up an ethnographic study capturing the everyday practices of lines on Istria. This involves recognition of how multiple borders can be at play at once, and that bordering practices may sometimes contradict and even conflict when enacted in the same space and time. In other words: "[T]he aim of border studies must be to decode borders, not as static entities bound up with specific self-evident concepts but as mobile, travelling borders, enacted in real time with all the contradictions and conflicts this may entail" (Andersen 2012, p. 144). The fieldwork upon which the chapter draws was conducted in shorter periods ranging from October 2009 to January 2011 as well as a more recent follow-up fieldtrip in July and August 2020. All interview citations in the chapter are anonymized and derived from the fieldwork. Initially the ethnography was not set up to explore the anthology's thesis on borderlands resilience, yet, returning to the results of previous fieldwork proved such an argument possible, and the follow-up trip was minded explicitly at researching the thesis.

Questioning the ownership of lines

Geopolitical borders never seem to settle on Istria. This is not just illustrated by historical circumstances, but it is also very much part of contemporary life on the peninsula. The dissolution of socialist Yugoslavia in the early 1990s inferred that what had been an administrative border between the two republics, Slovenia and Croatia, became an officially recognized state-border. Moreover, the state border between Italy and Yugoslavia went from being a border between two ideological systems to being "merely" a border between two independent nation-states. More recently, borders reiterated with the accession of first Slovenia and then Croatia into the European Union, making first the Slovenian-Italian state border and then the Slovenian-Croatian state border into internal EU borders – at the same time as the Slovenian-Croatian state-border became first an external EU, then external Schengen border (Krnel-Umek 2005).

This mapping is complicated even more so by the fact that the agreements in the 1990s on how to draw the borders of the new independent states stipulated that the administrative borders of socialist Yugoslavia were to be recognized as state borders. Many of these, including the Croatian-Slovenian border, were never formally drawn in the context of the second Yugoslavia (Krnel-Umek 2005). This lack of official border drawing has left many unresolved issues over borders on the Western Balkans causing ongoing re-settling of maps in the region. To mention just a few of these conflicts, there is an ongoing debate over whether the Vukovar Island in the Danube belongs to Croatia or Serbia; and the Neum corridor, separating the Neretva-Dubrovnik region from the rest of Croatia, remains a topic of dispute.

Istria is also influenced by ongoing border disputes since the break-up of Yugoslavia, illustrated most explicitly by the dispute over the Piran bay, beginning in 2008 and still not settled but only "parked" in an international committee. As part of the accession process, Slovenia demanded of Croatia to give up parts of its sea and coastal areas on Istria if it were to become a member of the European Union, thus forcing Croatia to enter renegotiation of the border. The EU helped join an international commission to keep the issue separate from the Croatian accession negotiations, but the commission has so far been unable to complete its task. Because of a corruption scandal on the side of the Slovenians in the commission, Croatia pulled out of the arbitration agreement in 2015 (Total Slovenia News 2020), something that is now also part of what is blocking Croatia's entry into the Schengen area.

Istria is in other words an area where traces of state borders are very hard to neglect by the local populations. The unsettled traces of lines involve many combatting claims for ownership of lines, a line-combat, which also constitute the larger region as "frontier society":

> [T]he dynamics of conflict in a frontier society have to be led back to the conditions of insecurity and violence that overcome all group divisions, creating a sort of existential instability for everyone . . . where the everyday fight for survival is made harsher by the continuous antagonism promoted by distant political centers.
>
> (Cocco 2010, p. 11)

This "unsettledness" constitutes an opening for constant discursive questioning of lines; years of experience have shown that lines do not settle, but that they are up for grasp and they can change. When Green uses the metaphor of "tidemarks" and reads borders as multiple lines, one on top of the other, it is indicative of such reflective opening. Wandji (2019, pp. 292–293) also points towards its importance in frontier societies when he says: "The border as a statement of external pressure began the discursive transformation of their geographical space, thereby shaping all other ensuing forms of adversity."

On Istria, "the unsettledness of borders" is fundamental for the everyday practices of lines. On the one hand, it enables a local distance towards geopolitics and an ability for the local populations to negotiate borders beyond and in a space different than official politics, negotiations that were often present during the fieldwork. For instance, when the Slovenian foreign minister raised concerns about the many open bars and nightclubs in the neighboring country during the summer of 2020, fearing that Slovenian tourists would return from their holidays infected with COVID-19 (Croatia News 2020), and when Slovenia responded to the pandemic not by denying Slovenians access to Croatia, but by closing the borders the other way and thus for Croatians, this was met with disbelief and scorning remarks in the Istrian context. As a waiter in a restaurant near the Croatian-Slovenian border said while complaining about the Slovenian decision not to let Croatians enter: "Half of Slovenia come here on holiday, but I cannot go and buy groceries in Koper [the biggest Slovenian town near the border]; if I do that, then it is a problem" (Conversation F). Or as the person behind the desk in the tourist information office told me, "It's political" (Conversation D), indicating how the locals understand disputes over lines as belonging to the space of the map made by state powers, a space that in their interpretation has nothing to do with real life on the peninsula. In the popular version of borders on Istria, they appear as ideological constructs whose meaning and importance are unfolding very far from everyday life on the peninsula: in Zagreb, in Brussels, in Ljubljana (Andersen 2012, p. 153).

Hence, when borders are debated on Istria it is often a question of distancing oneself from those who are responsible for making them, that is, those in power causing conflict, who do not have any understanding of life at the peninsula. Such anti-state sentiments equip locals to critically engage with lines of division. Lines are here read as were they done by "others," sentiments similar to what we see in other areas in the world such as the Italian south (Elbek 2021) and Northern Ireland (Andersen 2020). This critical distance towards line-making practice leaves a space for peaceful coexistence beyond the bordered space; conflicts are not here but there, not ours but theirs.

Yet, on the other hand, there is a maybe more cunning aspect to the distancing: It involves the constant mimicry of political powers and the lines communicated by them. Unlike in places where the reiteration of lines might be more sincere, the peninsula is influenced by the spirit of "playing along" for the sake of survival and personal benefit. As Cocco describes it:

> Consequently, the facts of lying, changing allegiance and converting can be explained in terms of strategies of preservation based on the necessity to maintain low profile and to show loyalty to the political authority and thus to secure normal existence.
>
> (2010, p. 11)

Hence the relation to lines-in-the-sand may seem untruthful, and "they are often considered as 'masks' that can be accepted or contested depending on political conditions" (Ibid., p. 11). Unsettled borders thus open for an ability to distance oneself from state bordering practice, while also playing with lines, a playing along with that in turn challenges lines and implicitly redraws them. Identifications and their related borderings are instituted and lived on Istria by distancing oneself from "the ownership of lines" and partly by taking control of lines and "playing along with" those who own the right to do lines – in other words, a "mimicry of power" (Ibid.).

Remembering imperial spaces

As mentioned earlier, the term "trace" indicates a sense of time in a way that the geometrical line normally associated with geopolitical borders does not. Read as trace, the line is not just cutting through space; it is referring to a past at the same time as it is present as line in the everyday life of the-here-and-now. Exposing lines in everyday life is therefore always-already more than snapshot exposures; it is also a matter of recognizing how "border logics" (Green 2012) reappear in everyday life by leaving marks – or "tidemarks" as Green calls them (2018). These logics mark contemporary ways of doing things, thus unfolding in multiple "new" practices, constantly reawakening the border of old, returning it to life, so to speak.

In the Istrian context, tidemarks articulate in not just nation-state histories and conflicts, but in memories of the past where the Austrian-Hungarian empire as well as Rome and Venice played important roles. Part of this (re) interpretation of the imperial past involves a more general learning process through which local populations have become aware of the destructive power of official border drawing, and the need for resilience in the face of change. As described by Wandji: "Colonial spatial delineation and the dynamics of colonial imperial competition destabilised local dynamics through imposing a reshuffling of geographical representation and spatial organisation" (2019, p. 292). However, unlike the lines related to state-drawing – especially those made over the last 30 years – the presence of empire is, on Istria, not just read and interpreted as power-struggles without much relation to the everyday life of people on the peninsula. Rather, empire is interpreted as mimetically resembling power and thus having real influence on everyday life, and it is relived in a creative reiteration of imperial lines. As Cocco says: "Here the myth of the imperial frontier acts as a powerful imaginary tool for the Istrian consciousness, reinforcing the conviction that Istria is a unique place for its bright past of imperial crossroad" (2010, p. 18). Without the presence of empires and all the struggles entailed, life on the peninsula would not have been what it is today.

Accordingly, (hi)stories of the imperial past provides a positive image of lines, much more positive than the lines made in geopolitical struggles over the last 200 years. The Austrian-Hungarian past is represented as "the golden period" with social tolerance, administrative efficiency and the ability to maintain order, as well as the fact of being an independent administrative unit in much of this period:

> Well, at the time when Austria ruled, we must confess, and now people agree, that the situation was far better. A few years ago, there were people alive, who still remembered Austria and they said that there was an impressive order, impossible to imagine today, that everything was working properly. . . . However, I think that this is also a myth that Croatia uses against Italy, against Venice, saying that we were Austrians, we were fighting on the same side (79, 2001, Opatija).
>
> (local inhabitant quoted in Cocco 2010, p. 19)

What might in other places in the Western Balkans be interpreted as Yugonostalgia is more of an Austria-nostalgia in these parts, and it is a nostalgia that is to a certain extent influenced by the antagonistic relation to the presence of "the Italian other" and a certain disorder and anarchy with which this other is associated. As one Croatian policeman controlling the border to Slovenia illustrates: "Italians just pass by the ques and try to cross the border avoiding the controls, not recognizing the order imposed" (Conversation B).

Yet, the Roman and Venetian influences are not just criticized and denied; on the contrary, they are recognized as culturally important. They concern the architecture, the eating habits on the peninsula, as well as the linguistic landscape on a peninsula where many people are multilingual, and one hears (Serbo)Croatian and Slovenian mixed with Italian. Moreover, as the Italian presence on the peninsula is read less as imperial presence then as a civilizational and republican presence, it does not seem to matter much that these imperial lines communicate something of an antagonistic relation: Italy and creativity/chaos vs. Austria and structure/order. What appear to be important are the ways these tidemarks appear and disappear, leaving space for each other and thereby working together to stabilize space in different civilizing forces.

Research on the Western Balkans has for some time now emphasized the local ability of "frontier societies" to creatively reinvent and re-settle in the face of adversity, as is the case in other parts of the world:

> However, precolonial communities, whose living spaces experienced a depression through colonial territorial demarcations, moved to sustain their status quo *ante*. These communities deployed many strategies to tolerate or circumvent the presence of imposed borders, drawing heavily on the colonial border's extraversion and weak structure to maintain their livelihoods, communal ties, and social organisation.
>
> (Wandji 2019, p. 292 following Lukong 2011 and Zartman 1985)

The "many strategies to tolerate and circumvent the presence of imposed borders" are an integral part of how tidemarks appear, and lines are enacted in everyday life in contemporary Istria, including an impressive ability to articulate imperial lines as recurring tidemarks, despite apparent contradictions and conflicts involved in doing so.

Intertwining national and regional lines

The smooth recurrence of tidemarks becomes somewhat difficult, however, when the tidal waves hit 19th- and 20th-century stories, where national lines take precedence. During my fieldwork when asking locals, "What is the status of national identifications in this borderland?" the answer I got was almost always: "Well, it is complicated." Reading national identifications depends, as we know, on how ethnicity is interpreted – as exclusive, inclusive, or as unimportant (Jenkins 1997) – and in the Istrian context all possibilities are at play at once, making for a hard-to-read-complexity and for subtle situatedness and flexibility of line-making (see also Andersen 2012).

If we begin in one national context, it has been well documented by now that since the 1990s, the identification with the Croatian nation was fuelled by ethno-nationalism, and thus use of ethnicity at political level to legitimize the right to power in the Croatian territory (Pavlakovic 2014). Independence and national sovereignty have been interpreted as the "real common good," integrating all legitimate positions in the political spectrum with a common struggle for Croatian sovereignty and making ethnicity the most important indicator of political loyalty. Positions other than this were illegitimate, however innocent, and non-aggressive they might be, making it very difficult for Istrians to identify with the new Croatian nation. During my fieldwork, most Istrians would deny allegiance to this type of ethnicity and identity, not the least because of their understanding that line-practice is always-already unsettled; as illustrated, lines of division change, and this obviously exclude the belief in partition as a dream come true.

Turning to the other major national identification influencing the peninsula, the Italian, the line-practice is yet another because this ethnic affiliation locates in lines of violence and ethnic cleansing, rather than victory. After having been an autonomous administrative district from 1939 to 1954, the larger part of the peninsula became part of socialist Yugoslavia (Zona A remained Italian, but Zona B went to Yugoslavia). Most Italians left the peninsula, approximating 200,000, and it remains disputed if Yugoslavia carried out ethnic cleansing forcing these people to leave, or if they chose to do so on their own account. What happened to people at this point in time is not much spoken about, leaving lines of tension on the peninsula, not just between Croatians/Slovenians and Italians but also between Italians on the peninsula; on the one hand those who call themselves "exiles" and claim they were forced to leave and, on the other, the Italians who stayed behind, claiming no one was forced to leave.

Yet despite of knowing about this past and its connotations, Istrians do not seem to deny their Croatian-ness or yearn for a lost Italian-ness. Rather, they practice nationality in a regional appropriation of both (and more), which brings us back to a dimension of line-practice mentioned previously, namely that of mimicry and play. When lines are understood as what they are, lines on a map, then they become open for play. In terms of identifications, it opens the kind of performativity where the actor fits his or her line-practices to the situation without aiming for the stability of a single identity. This could involve being Croatian in the morning and Italian in the evening after having had a few drinks (Drakulic 1996, p. 163). A mimetic identification is carried out to fit into whatever the circumstances ask for. Such "masking" must be understood in the context of an everyday struggle for survival, performed as a mimicry of power structures and recognizing them as ambiguous games (Cocco 2010). The distance between reality and map, between the line's ontological and epistemological status, is an opening that is used (and abused), constantly creating new tidemarks, all of which are understood to resemble reality as well as mimic it.

The fact of being able to play with many possibilities – the multi- of any identification – sometimes reiterates into a certain "Istrian-ness":

> While one's nationality could change, there would be a permanent pattern of identification, namely the local identification with the territory. This identification, synthesised by the concept of Istria-ness [istrijanstvo] would mean multi-ethnic and multi-national sense of belonging to the territory of a frontier region, beyond state borders – as it is specified in the statute of the regional administrative body.
>
> (Cocco 2010, p. 23)

When a museum curator in Umag talked with me about local identifications, she told me that the "Istrian identity is geographical" (Conversation E), emphasizing how the most important identification on the peninsula is with the location, its landscape, customs and history.

Paradoxically, the Croatian national attempts at instituting lines to part nationalities has actually helped emphasize an Istrian regional ability to live a life beyond such dividing lines:

> [. . .] the ethnic politics that privileged the "Great Croats" of the diasporas also increased a sense of insecurity, which had a territorial projection. In other words, the state sponsored nationalist discourse eventually strengthened the regional differentiation that emerged in the public sphere after the work of some local elites, already in the eighties.
>
> (Cocco 2010, p. 21)

This form of resilience where locals identify with and take control over regional space to avoid ethno-nationalism is witnessed all over the Western Balkans (cf. Hudson and Bowman 2012); however, on Istria the peninsula's peripheral status and years of experience might have given its populations an advantage compared to other localities in the former Yugoslavia.

Juggling multiple lines

Appropriating Massey's notion of "a simultaneity of stories-so-far" (2005, p. 12), a concept that captures how together different times, practices, aspirations and failures condition the possibilities of future practices, Green points out how tide-marks express borders at once as physical lines ordering the messiness of everyday life; as stories told to reveal or cover up previous marks; and as this multiplicity at once:

> Once an official border has been built, using some kind of classificatory logic, many worlds could be generated thereafter, and not just one. I have limited my discussion to dominant, singular and formal border logics for the sake of clarity and brevity; but the obvious point that borderness can be multiple, even to the extent of some people recognizing a place or a thing as a border while others do not see anything except landscape, is a crucial aspect of what could be called borderness dynamics for lack of a more elegant phrase.
>
> (2012, p. 581)

Lines are not endowed with uniform meaning but endlessly (re)defined, and "[b]order-ness concerns where things have got to so far, in the multiple, unpredictable, power-inflected, imagined, overlapping, and visceral way in which everyday life tends to occur" (Green 2018, p. 81). Lines are in other words always "under construction," they are open for "play," and there is no one way of appropriating them.

In the line-practice of Istrians, this also involves the ability to "play" with identifications beyond apparent contradictions and conflicts. As was the case with the Austrian-Italian heritage, the obvious conflict and ongoing struggle between national and regional lines on Istria does not develop into two irreconcilable identifications unable to find common ground, but rather opens towards a productivity of the layers; that is, the many ways in which lines are articulated come to express multiple borders (cf. Andersen 2012). Lines change in time, and these rearticulations are also recognized as needed for their appearance in the real-time. As Massey says:

> [A]ll borders are multiple, generated from multiple vantage points – though of course, this does not mean that people are free to imagine border in any way they please: the simultaneity of-stories-so-far, and the entanglement of relationships and "power geometries of space" regularly constrain whatever vantage point emerges.
>
> (Massey 2005, p. 16)

Hence, the identifications appealed to cannot be read as merely being a "micro-world" resolving the tensions between national belonging and political affiliation. Rather, it should be understood in terms of an active reiteration of lines, where the interference between ethnicity and nationality with its claim of purity and exclusiveness no longer constitutes a hegemonic force to be either applied or denied. In this landscape of multiple lines, blood relations matter less than the soil upon which you live, the land you inherited from your ancestors. Struggles will continue over this land and, following the logic of imperial lines, the soil is therefore made up of multiplicity (thereto, see also Frandsen 2022, p. 124ff.). As Cocco reminds us:

> Therefore, all reflections on the sense of belonging in South Eastern Europe should consider the multiple dimension of this cultural space where the ethnic dimension is only a convergence of some dimension of identity, which nevertheless crosscut and spill over the rigid ethnic border.
>
> (2010, p. 9)

It is because of the crosscutting and the related uncertainty and unsettledness that identifications can appear multi-, anti- and beyond, all at the same time, if needed. As stated by the curator in Umag mentioned previously: "People deal with tensions because they are used to them. They never know what will happen next in term of geopolitical changes" (Conversation E). There is a historical memory of uncertainty in these parts, and people have learned to live with it, also by living beyond conflicts, which would influence most other populations confronted with similar problems (cf. Wandji 2019; Ferdoush 2022). Furthermore, research shows how line-practice involved in people's identifications influences how they locate themselves in their social worlds and how they form relationships with others (on such boundary work, see Lamont and Molnar 2002), and Istria is no exception. The multiple memories of lines to be evoked in everyday life are making social space here, as in how people move and talk about the place, making and limiting "spaces of being and becoming" for themselves and others, thereby also forming enclaves and other habitats (for similar social dynamics, see Lamour and Blanchemanche 2022). A resident in Bibali, a village situated close to the Croatian-Slovenian border, told me: "My father has gone to work in Trieste for 15 years now" (Conversation C), implicitly saying that it is not a problem for locals to cross the borders in their everyday life. Lines are lived with and in accordance with manifold others who are part of the peninsula because of its history; they are lived both with and in spite of the nationalists who say that there can be only one line; and they are lived beyond the space of lines, enacting them only when it truly matters.

Therefore, the constant renegotiation and search for protection expressed in these line-practices do not necessarily constitute problems. Here it is not the settling of lines, nor the neglect of conversations about lines, that is integrating and indicative of resilience. Rather, it is the constant rearticulation of lines in the everyday lives of people on the peninsula, which makes the place resilient towards conflict escalation. Locals themselves practice and revitalize the ability to deal with ontological insecurities and the tensions deriving from them, thereby redrawing lines not to separate, but to interact in ways where multiple identities are respected and valued (similar to the notion of "sensitive spaces" referred in

Ferdoush 2022, p. 108). This obviously does not make lines of division unimportant; it intensifies the importance of borders because on Istria identities do not come without them.

Conclusions

During my fieldwork on Istria, the chief of the Croatian border police on Istria made the statement that "borders are all in the head, and when we realize that, we can also diminish their importance" (Conversation A). The chief of police recognizes that borders are epistemological constructs drawn on maps and thus "only" separating imagined communities (Anderson 1989). What the utterance exposes us to is not a denial of the existence and importance of borders; it is indeed the very opposite. Despite of being "in the head," borders are indeed very real; they obviously matter in people's lives and this in powerful ways, also for the chief of the border police, who is employed to guard them and thus earn his living because of their existence. Rather than denying the importance of borders, the utterance exposes us to a productive, reflective leap between the epistemological presence and the ontological absence of borders, which prevents any drawing of lines in the sand from finally being settled (cf. Massey 2005; Green 2012, 2018). The chief of police appears to understands how the reflective leap works or, with Tomizza's words quoted at the beginning of the chapter, the chief of police "identifies with the border," thus knowing how to "play by its rules."

In Istria, it is the reflective leap between the line on the map and the line in reality that allows intersections between geopolitics and everyday life to express in a landscape of lines, where resilience towards external stress is first and foremost one of coping with geopolitical transformations imposed by powers located far away from the peninsula. Guiding life at the peninsula are images and memories of lines making it possible to (re)appropriate borders and divisions in multiple ways and as part of everyday life while actively engaging in (re)drawing and erasing lines. This entails a constant renegotiation of what it is to be Istrian. The lines in the sand truly matter on the peninsula, not in terms of their ability to divide but in terms of their ability to open towards multiplicity and ultimately the play of difference and belonging, which is much more of a staged performance here, than a belief in the integrity of the hegemonic, single identity.

Some go as far as to claim that this multiplicity equals the regional line-practice, also constituting one of the most attractive features of life on the peninsula. Adverts made to attract tourists thus provide invitations to think romantically about Istrian identifications in their multiplicity as the normatively correct way of performing identifications and living relations to others. As the curator in Umag put it: "Istria is the only part of Croatia which is European," inferring that Istria is the only part of Croatia where the European motto of United in Diversity has been truly implemented; peninsula inhabitants are indeed regionally united in their national diversities. Some Istrians even understand the peninsulas' inhabitants to live according to a cosmopolitan disposition enabling endless inclusion of otherness into local identifications. The inhabitants are thus living out the relation to other in what would for many be the normatively most appropriate way.

However, such romanticism clearly involves forgetfulness of the reality of the situatedness, that is, the stage upon which the multiple lines are played. As Cocco reminds us:

> The nature of such borderland identity . . . reveals a function though: the personal search for protection and even physical survival during dangerous transition process that was taking place in an area where state identity is frequently subject to changes and nationality is often a matter of discrimination.
>
> (2010, p. 23)

Living lines in "the Istrian way" is part of a landscape constituted by geopolitical instability, by everyday struggle for survival and by feelings of being constantly under siege. Moreover, the performances are often spiteful and expressive of hatred of those who are bound to more stable identifications, as much as they are self-indulgent and cunningly aware of their value at the market of dreams when played out and performed for the highest bidder. The Istrian experience is thus an altogether efficient form of resilience in times of precarity, yet the experience would be stressful for those who are not used to it and therefore not necessarily the most attractive form of life ready to be promoted universally and for all.

References

Andersen, D.J. and Sandberg, M., 2012. Introduction. *In:* D.J. Andersen, M. Klatt and M. Sandberg, eds. *The border multiple: the practicing of borders between public policy and everyday life in a re-scaling Europe*. Aldershot: Ashgate, 1–22.

Andersen, D.J., 2012. The multiple politics of borders. *In:* D.J. Andersen, M. Klatt, and M. Sandberg, eds. *The border multiple: the practicing of borders between public policy and everyday life in a re-scaling Europe*. Aldershot: Ashgate, 141–159.

Andersen, D.J., 2020. Gensyn med Ballybogoin. *Politica*, 52 (4), 420–437.

Andersen, D.J. and Prokkola, E-K., 2022. Introduction: embedding borderlands resilience. *In:* D.J. Andersen and E-K. Prokkola, eds. *Borderlands resilience: transitions, adaptation, and resistance at borders*. London: Routledge, 1–18.

Anderson, B., 1989. *Imagined communities*. London: Verso.

Cocco, E., 2010. Borderlands mimicry: imperial legacies, national stands and regional identity in Croatian Istria after the nineties. *Nar umjet*, 47 (1), 7–28.

Crang and Travlou., 2001. The city and topologies of memory. *Environment and Planning D: Society and Space*, 19 (2), 161–177.

Croatia News., 2020. Slovenia to decide on possible restrictions on arrivals from Croatia. Available from: www.croatiaweek.com/slovenia-to-decide-on-possible-restrictions-on-arrivals-from-croatia/ (accessed 4 January 2021).

Drakulic, S., 1996. *Café Europa*. London: Penguin Books.

Dürrschmidt, J., 2002. 'They're worse off than us' – the social construction of European space and boundaries in the German/Polish Twin-City Guben-Gubin. *Identities: Global Studies in Culture and Power*, 9 (2), 123–150.

Elbek, L.L., 2021. Rupture, reproduction, and the state: the Arab spring on Lampedusa as 'layered event'. *History and Anthropology*, 1–18.

Ferdoush, M.A., 2022. Stateless' yet resilient: refusal, disruption and movement along the border of Bangladesh and India. *In:* D.J. Andersen and E-K. Prokkola, eds. *Borderlands resilience: transitions, adaptation, and resistance at borders*. London: Routledge, 106–118.

Frandsen, S.B., 2022. Schleswig. From a land-in-between to a national borderland. *In:* D.J. Andersen and E-K. Prokkola, eds. *Borderlands resilience: transitions, adaptation, and resistance at borders*. London: Routledge, 121–136.

Green, S., 2012. A sense of border. *In:* D. Wilson and H. Donnan, eds. *A companion to border studies*. London: Wiley, 173–189.

Green, S., 2018. Lines, traces and tidemarks. *In:* O. Demetriou and R. Dimova, eds. *The political materialities of borders: new theoretical directions*, Manchester: Manchester University Press, 67–83.

Hudson, R. and Bowman, G., 2012. *After Yugoslavia. Identities and politics within the successor states*. Basingstoke: Palgrave Macmillan.

Jenkins, R., 1997. *Rethinking ethnicity: arguments and explorations*. New York: Sage Publications.

Krnel-Umek, D., 2005. *The Slovenian-Croatian border in Istria – past and present*. Ljubjana: Narodna univerzitetna Knijznica.

Lamont, M. and Molnár, V., 2002. The study of boundaries in the social sciences. *Annual Review of Sociology*, 28, 167–195.

Lamour, C. and Blanchemanche, P., 2022. A resilient *Bel Paese*? investigating an Italian diasporic translocality between France and Luxembourg. *In:* D.J. Andersen and E-K. Prokkola, eds. *Borderlands resilience: transitions, adaptation, and resistance at borders*. London: Routledge, 152–156.

Lampe, J.R., 2000. *Yugoslavia as history. Twice there was a country*. 2nd edition. Cambridge: Cambridge University Press.

Lukong, H.V., 2011. *The Cameroon-Nigeria border dispute. Management and resolution, 1981–2011*. Oxford: African Books Collective.

Massey, D., 2005. *For space*. London: Sage Publications.

Parker, N. and Vaughan-Williams, N., 2009. Lines in the sand? towards an agenda for critical border studies. *Geopolitics*, 14 (3), 582–587.

Pavlakovic, V., 2014. Fulfilling the thousand year-old-dream: strategies of symbolic nation-building in croatia. *In:* P. Kolstø, ed. *Strategies of symbolic nation-building in South-Eastern Europe*. London: Routledge, 19–50.

Rumford, C., 2008. Introduction: citizens and borderwork in Europe. *Space and Polity*, 12 (1), 1–12.

Taussig, M., 1993. *Mimesis and alterity: a particular history of the senses*. London: Routledge.

Total Slovenia News., 2020. Decades old border dispute between Slovenia/Croatia still unresolved. Available from: www.total-slovenia-news.com/politics/6381-decades-old-border-dispute-between-slovenia-croatia-still-unresolved-feature (accessed 3 January 2021).

van Houtum, H, Kramsch, O. and Zierhofer, F. eds., 2005. *B/ordering space*. Aldershot: Ashgate.

Wandji, G., 2019. Rethinking the time and space of resilience beyond the west: an example of the post-colonial border. *Resilience: International Policies, Practices and Discourses* 7 (3), 288–303.

Zartman, W., 1985. *Ripe for resolution: conflict and intervention in Africa*. Oxford: Oxford University Press.

List of sources

Conversation A: Istria Border Police HQ, 15.08.2011.
Conversation B: Plovanija border check point, 23.08.2011.
Conversation C: Bibali, 27.07.2020.
Conversation D: Tourist Office Umag, 30.07.2020.
Conversation E: Umag City Museum, 30.07.2020.
Conversation F: Restaurant Umag, 03.08.2020.

12 Epilogue

Borderland resilience: thriving in adversity?

Jussi P. Laine

Borderlands around the globe have been undergoing considerable transitions due to the ever-escalating trend of tightening control at state borders. We have witnessed a consistent drive for ever-stricter border and migration policies that have not only changed the role borders play but become inherent parts of a wide range of polices and societal practices. Indeed, despite many appealing aspirations to the contrary, political borders have proven their persistence. The common response to the geopolitical global tensions, broad societal challenges and multiple overlapping crises has been to regress to state-centric thinking and nationalist agendas, and to ad hoc border closures (Laine 2021). In Europe, in particular, the witnessed regression into nationalistic, state-centric thinking and the commonplace populist reduction of borders to mere protective frontlines overshadows the innovative conceptual (re)framing of borders as social, political, economic and cultural spaces and resources.

Recent events around the globe and the related frequently reported expressions of neo-nationalism, populism and xenophobia, as well as border violence, may appear to refute the potential of borders to connect. Amidst the uncertainties and insecurities of the current era, accentuated further by the current COVID-19 pandemic, the role of borders as interfaces between domestic and foreign concerns, as markers of difference and as barriers to undesirable influences and foreign threats has only been reinforced (Laine 2020a). However, these developments can be viewed as testimony to the suggested wider social relevance of borders – and it is herein that border studies as a multidisciplinary – even post-disciplinary – field of investigation has the potential to make a difference and help us to better understand and interpret the complex transformations that our societies are facing. As much of the recent border studies literature indicates, borders are being constantly reconfigured as political arenas, sites of contestation and spaces of possibility – all at the same time.

Borders have acquired a central position in the social and political transformation of the world, having a substantial impact on the daily lives of many people. Rather than mere independent variables in the analysis of political processes, borders are co-constitutive of political agency: borders are not given but emerge through socio-political and cultural bordering processes that take place within society (Scott 2020, p. 4). Borders connect the different political, social,

DOI: 10.4324/9781003131328-15

economic, technological, cultural and psychological processes in understanding how the world works processes in space, in time and at the level of everyday life. Yet dynamic processes and phenomena, as well as responses to them in and across borderlands, remain insufficiently understood and represented. There is thus an apparent need to examine how people in border areas actively engage in making and sustaining cross-border relationships, as these interactions shape life in the region, in the process reconfiguring cultural and political landscapes (Ptak et al. 2020). After all, "cooperation across scales and times forms an essential factor for human adaptability as connections and networks across borders facilitate social interaction, the flow of ideas and resources" (Andersen and Prokkola 2022, p. 2; Davoudi et al. 2013; Korhonen et al. 2021). At the same time, the disruptive forces of change – whether real or imagined – elucidate the main argument of border studies: that borders are in a constant process of confirmation, contestation, transformation and re-confirmation (Scott 2020, p. 4).

The various dynamics of globalization have profoundly changed the power of borders, modifying the dialectical relationship between their fixed institutional nature and the constantly changing processes of bordering within and between societies (Konrad 2015; Rumford 2012). However, the observed changes cannot be credited to the broad globalizing tendencies only; from the perspective of the borderlands, the dialectical relations may have been transforming for a long time, whereby borderlands resilience express everyday enactment of identifications and historical memory in borderlands (Andersen 2022). Borders are sites of encounter and interaction, but also manifestations, performances and representations of territorial identities and biopolitical control. They are constantly reconstructed and maintained as frames of social and political action, strategies of challenge, survival and the related patterns of identification and identity politics, as well as symbolic social and cultural lines of inclusion, encounter, difference and contestation (Andersen et al. 2012). While various conceptualizations and interpretative tools have been proposed to better capture the ever-evolving multiplicity of borders, there seems at least to be a strong academic consensus about the inherent complexity of the border concept. In "a world at once more bounded and more borderless" (Konrad et al. 2019, p. 2), however, it is precisely this complexity of borders that has fuelled a general tendency to simplify the complex realities into virtually opposing imaginaries of either a borderless or a nationally segmented container world that are constantly reproducing themselves through each other (see Paasi et al. 2019; Kraemer 2020; Laine 2021).

When investigating border complexity, we must also consider ideational counter-currents: imaginaries of borders as politically relevant frames of reference that function based on simplifications of complex situations and the confirmation of received notions about the (social) world. Such imaginaries contribute to a stabilization of identity, but also involve potential traps of static thinking: an imagined familiarity that can prevent more dynamic understandings of the world. It is self-evident that the politics of simplification and selective visibility do little to recognize the plurality of the voices that operate in everyday life – politically, socially and culturally. Making some aspects of border complexity irrelevant or socially non-existent contributes to the construction of social imaginaries through the simplification, or even obfuscation, of wider social, economic and political dynamics.

The present volume makes a valuable contribution to our border knowledge by clarifying the relationship with and relevance of borders to transnational and transcultural phenomena at various levels, as well as their embeddedness in people's everyday relationships and interactions. Introducing the resilience perspective in studies of borders and borderlands allows the authors to assess how diverse groups of people whose daily lives are entangled with borders and border crossings maintain wellbeing and adaptive capacities in the face of border transitions, be it reinforced securitization or new openings and opportunities to transgress contemporary state b/ordering. The authors challenge the view that the stable and fixed boundaries modernity has commonly seen to have established would have naturalized borders as rigid lines that not only divide communities but also determine their future. In such modernist ontology, insecurities have come to be managed by control and order. Fluxes and flows have been reduced to fixed quantifiable forms and entities, complex mobile phenomena into rigid Cartesian confines that are knowable, governable and manageable by people with the powers of reason. Confining space by bordering out the unpredictable has brought psychological comfort and enforced the illusion that we are indeed in charge (Laine 2020a). In seeking independence, salvation and security, the modern subject of man [*sic*] has effectively distanced itself from the rest, of which it is utterly part and contingent, in so doing also making itself increasingly fragile and vulnerable.

However, the markings of the power relations, hierarchies, exclusions and prejudices, as well as the entire international system that maintains them, have come under pressure amidst the inescapable effects of increased global interconnectivity, but at the same time by more situated on-the-ground, everyday borderland realities. While much of resilience research and interventions tend to take bounded regions and communities as a point of departure (Healy and Bristow 2019), the present imaginaries of risk or threat seem now hardly spatially and temporally bounded; rather, the unexpected feedback loops and side-effects of our actions have made the threats increasingly unpredictable and uncontrollable. They cannot be mitigated by enforcing and relying on the awareness that we could prevent them if only we wanted. Apart from the impacts of the perceived risks in themselves, the shattering of the very foundations of the order and logic, no matter how subjective or biased, to which one may have become accustomed can in itself be seen as a cause of insecurity.

Whereas the past modernist logic insinuated that fears had to be contained to maintain local and global order, the augmented unboundedness and fluidity of the current era have not only blurred the foundational division between "us" and "them," but also complicated the very criteria based on which these binary groupings used to be defined. That is, the augmented unboundedness comes with benefits, but may also fuel fears of fading hierarchical taxonomies and generate illusions of invasion and terror. The anxieties caused by the widespread fear that all difference will be eliminated, reducing everyone to the same level of vulnerability and erasing all the distinctions legitimizing notions of belonging, identity and privilege (Giuliani 2021) may easily become calls for tighter borders, military violence against the "enemy," thicker collective identities and more inequality to preserve the "we" (Latour 2018).

In a world beyond control, bordering the uncontrollable seems both inefficient and counterproductive. Rather than seeking common solutions to challenges that are inherently international, many states have reverted to national solutions or short-sighted border closures. Post-9/11 organized terrorism, narcotics trafficking, unruly human mobility from the Global South to the Global North, the ever-evolving environmental crisis – as well as the ongoing COVID-19 pandemic – have unleashed a complex assemblage of anxieties, fears and apocalyptic discourses that have gained even more presence by having been differentially reproduced by various national and international actors involved in, but not limited to, border control and management (Giuliani 2021; Laine 2020a). While increased investment in border security may create an impression that the threats are being taken seriously and something is being done to keep them away, borders in the big picture appear the problem rather than solutions to what is at stake. Thus, rather than bordering the threats out and insulating us in, academics, practitioners and policymakers alike have turned their attention to the concept of resilience; that is, from the need to merely respond to a threat or fighting against it to building a holistic capacity to recover quickly from the faced challenges, maintain a positive adaptation despite experiences of significant adversity, and move on without succumbing to the pressures constant crises inflict on us.

As Andersen and Prokkola note in their Introduction, resilience thinking and research have indeed proliferated in various fields and disciplines. Resilience has gained momentum as a policy concept focusing on a highly diverse set of issues (Dunn Cavelty et al. 2015), and within various policy fields the response to a "world of rapid change, complexity and unexpected events" has been discussed (Chandler 2013, p. 1). While this has certainly advanced the ways we approach and understand uncertainty, it is equally evident that major discrepancies remain in conceptualizations of resilience, and little consensus seems to exist around the central terms used within the various models of resilience that seek to capture the significance of contingency, vulnerability, security and protection. As Humbert and Joseph (2019) contend, resilience is not a thing, nor is there "one" logic of resilience that informs all instances of its usage, yet this realization should not automatically be taken as a limitation for its use. As Brand and Jax (2007) suggest, resilience has actually become a "boundary object" that facilitates communication across disciplinary borders by creating a shared vocabulary, even if the understanding of the parties differs regarding the exact meaning of the term in question (see Star and Griesemer 1989). The potential thus lies in the conceptual vagueness and malleability of resilience (Brand and Jax 2007), which should be embraced by providing contextualized approaches that also consider the crisscrossing of various scales and bounded entities in contrast with more rigid "placeless" benchmarking measurements (Bristow 2010), which is exactly what the individual chapters in this volume have done. The resilience of "people of the borderlands" is situated in diverse contexts and can occur in response to such pluralities of disruption (Wandji 2019; Svensson and Balogh 2022).

Resilience thinking holds great potential – to which the present volume serves as testimony. The current COVID era has also accentuated the need the conceptualize resilience not merely as an ability to recover and bounce back as quickly as possible after disturbance, but rather as an ability to adapt, acquire new capabilities and emerge stronger from the struggle. Resilience is not about securing from danger, pre-emption or precaution, but about adaptive risk management (Bettini 2017, p. 89). Resilience is not, however, a universal magic bullet for a complex set of challenges or threats, nor is its capacity to deliver security irrefutable. In acknowledging that the recent critical resilience thinking has been excessively fixated on resilience as participating in a neoliberal rationality of governance (Mavelli 2019; Wandji 2019), the present volume extends readings of resilience by turning its gaze on a more quotidian conception of resilient activity at the borderlands – that is, on "people's resilience," by which the volume seeks to draw attention to different social groups' ability to self-organize and mobilize skills and resources to create opportunities when faced with adversity, and to act in solidarity when their community is disturbed and even disrupted (Andersen and Prokkola 2022). The chosen focus holds a great potential as social capital has been widely recognized as an enabling resource and playing a key role in building and maintaining social resilience (see e.g. Adger 2000).

In addition to advocating for the multi-scalarity of the resilience concept to be applied situationally, and thus be understood in *the* context, the key contribution that the present volume makes lies in the emphasis placed on the thus-far insufficiently utilized border(land) approach. As many border scholars and policymakers alike have recognized, most things that are important to the changing conditions of our societies and economies take place along borders (both international and internal), and some of these – for example migration, commerce, smuggling or security – can be found in borderlands in even sharper relief (Wilson and Donnan 2012). Borderlands have distinct features and unique characteristics due to either increased interaction or lack thereof. The imminence of the "other" has had a propound impact on the way in which borderlands have evolved and developed specific forms of living together that entail tolerance and solidarity. For the people living on either side of the border, for their lives and livelihood, and to the everyday practices of exchange and interaction, the role of the border – be it material, physical, symbolic or imaginative – is crucial, yet not necessarily the same for all. Borders are persistently made and remade, in concert with larger the projects of states, historical memories, circulations of migration, implementation of trade accords, and political responses of those living through and in these processes.

As the editors explain, the borderland approach provides a new understanding of the significance of the geopolitical environment, political contestation, and values in fostering borderlands resilience. In other words, borders provide us with an illuminating operational looking glass to observe a broad array of phenomena and forces of change that can strongly affect our societies and the borderland people. Perhaps most importantly, the volume provides an appealing and convincing argument that rather than being a source of security and a key factor in amplifying borderland resilience, borders often create stressful situations and only increase experiences of the disruption of daily life (e.g. Lois et al. 2022; Hannonen 2022). Indeed, the extent to which boosted border security or ad hoc border closures actually make people feel safer and our societies more resilient remains highly debatable (Laine 2020b).

Chandler (2019) goes as far as to argue that if understood in these modernist ways, resilience is itself part of the problem rather than the solution. Acknowledging that resilience thinking has recently achieved nearly universal success in the policymaking world, he posits that

> resilience is still a "modern" construction which assumes that problems are "external," and that we need to develop "internal" policy solutions to maintain and to enable our existing modes of being in the face of shocks and perturbations. "We" need to be more responsive and adaptable.
>
> (Chandler 2019, p. 305)

In my reading, the borderland perspective assumed here can be used to escape the clear-cut modernist *dispositifs* and dualisms that Chandler grasps. As the editors underline, it is important to recognize that resilience is not inherently bound to any particular bordered administrative unit; the view from borderlands complicates the prevailing territorially bounded understandings of social resilience. Inspired by Wandji's (2019) work, they call us to recognize that the diffuse resilience of communities dealing with the presence and persistence of externally imposed international borders provides material resources and practices for uncovering the border itself as a disruption rather than a protection. Resilience is indeed often described to stem from the fact that "the opening of state borders has made 'places and regions more permeable to the effects of what were previously thought to be external processes'" (Christopherson et al. 2010, p. 3; Prokkola 2021, p. 4). Reducing resilience to a neoliberal strategy ignores the complex mundane practices of people and communities (Wandji 2019), which is what the authors of the present volume seek to emphasize by questioning what counts as resilience.

Evidently, the top-down politics of resilience is interlinked with the current ambivalent border governance and security thinking, in which the perceived threat imaginaries stemming from a "borderless world" are often employed to explain the increasing importance of developing resilience (Prokkola 2022). Whether by showcasing the role of frictionless cross-border mobility as an important factor contributing to regional stability and resilience building (Koch 2022), underlining the need to include "the other side" in what it means to be "local" in fostering cross-border communities and borderland resilience beyond mere coping and adaptation (Svensson and Balogh 2022), showing how revitalizing minority languages can play an important role in the ability of borderland populations to mobilize identity-formation and histori-cal memories as a means of resilience (Ridanpää 2022), or investigating the role of diasporic trans-localities in the course of an intensified urbanization of the border area (Lamour and Blanchemanche 2022), the authors underscore that it is the various cross-border connections, rather than borders them-selves, that play a key role in fostering resilience. Still, we cannot deduce from this that borders in themselves would necessarily be problematic endeavors. Rather, as Andersen (2022) shows, borders are worked with, crossed and overcome, and sometimes even used strategically to articulate diversity in an otherwise integrated space.

Resilience ties security to insecurity (Dunn Cavelty et al. 2015, p. 11). It pro-vides a conceptual framework that re-orients thought and practice in response to the Anthropocene's destabilizing effects (Grove and Chandler 2017) and holds out the promise of living with, if not benefiting from, change, uncertainty and vulnerability (Aradau 2014; Evans and Reid 2014; Grove 2014). How, then, different subjects seek to establish resilience and what kinds of effects this produces – be it in terms of politics, society or ontology – may take various forms (Dunn Cavelty et al. 2015, p. 12; Laine 2020b). Efforts to resist transi-tions and uncertainties of border closures and securitization in the everyday lives in the borderlands are vividly exemplified by Hannonen's (2022) explo-ration into borderlands resilience. In discussing the local effects of national and international mobility developments, she emphasizes the role of human agency in resilience in a cross-border mobility context through the notion of co-construction of resilience through practices and responses. The mere border crossing, however, does not necessarily bring about the transgression of bor-ders. Rather, as Iwabuchi (2016) notes, it requires one to fundamentally ques-tion how borders in the existing form have been socio-historically constructed and seek to displace their exclusionary power that unevenly divide "us" and "them" and "here" and "there."

While the structural change in borderlands often loses sight of the nuances of daily life, Ferdoush (2022) depicts how the human agency may take shape in the forms of everyday practices of refusal of existing order as an integral part of survival and thus resilience along borders. Contrary to the increased materiality of borders that has become an integral part of the lives of borderlanders worldwide, mobility practices across borders can thus be seen as a refusal of the state superstructure and an indication of how, having perceived a disruption to their geographical imag-ining, borderlanders proceed to maintain connection between the social spaces divided by the border (Wandji 2019). The borderlands resilience has its own logic that is interconnected with yet simultaneously different from the higher-level agendas, which has become increasingly evident amidst the COVID-19 pandemic as the intensifying border closures and state securitization leaving has left the borderland populations to, yet again, reiterate their identities anew (Lois et al. 2022).

The common assumption in the resilience literature is to regard threats as something external (Humbert and Joseph 2019). A borderland perspective can, however, be used to the rethink this assumption, and even to challenge its core premise. As an example, Wandji (2019), whose work the editors of the volume have wisely chosen to guide their thinking, urges us to expand our epistemological approaches to resilience thinking. He reconfigures borders in terms of a "plurality of disruptions" to continuity, something that embraces the idea of the need to be resilient in the face of constant challenges, while at the same time questioning the idea that their source must always be external. This brings us to the need to address the concept of security by recognizing that it is not limited to inter-state relations or issues of a merely geopolitical nature, but is interwoven and "intermestic" where domestic and international concerns meet (Laine et al. 2021). Border communities must constantly reassess their geographical imagination following various historical developments and policy changes in the wider international context that affect them directly and adversely (Wandji 2019, p. 289). That is, border communities tend to be resilient essentially in terms of adaptation as a form of continuity rather than change. As Frandsen (2022) argues, the focus on borderlands is particularly useful when discussing the limitations of national historiographies, because it enables an understanding of continuities, flows and contacts that is independent of the institutions of the nation-state borders of a later date. Borderland resilience thus stems from the situated traditions and the multiplicity of experiences of living with and adapting to the dynamism of a border(land).

Resilience thinking underlines a comprehensive understanding of security in positing an entire system's capacity to spontaneously reorganize itself in response to disturbance, adapt and transform. Whereas politics is commonly reduced to responding to and managing what are understood as the consequences of previous human actions, assuming a transformative resilience-oriented approach can be regarded as a depoliticizing effort, because conditions of responsiveness, adaptability, inventiveness and flexibility are required to survive and prosper within a catastrophic horizon (Grove and Chandler 2017). A resilient world is a world of becoming. It necessitates the transcending of boundaries and the binaries of which they are markers.

References

Adger, W.N., 2000. Social and ecological resilience: are they related? *Progress in Human Geography*, 24 (3), 347–364.

Andersen, D.J., 2022. Line-practice as resilience strategy: the Istrian experience. *In:* D.J. Andersen and E-K. Prokkola, eds. *Borderlands resilience: transitions, adaptation, and resistance at borders.* London: Routledge, 166–181.

Andersen, D.J. and Prokkola, E-K., 2022. Introduction: embedding borderlands resilience. *In:* D.J. Andersen and E-K. Prokkola, eds. *Borderlands resilience: transitions, adaptation, and resistance at borders.* London: Routledge, 1–18.

Andersen, D.J., Klatt, M., and Sandberg, M., eds., 2012. *The border multiple: the practicing of borders between public policy and everyday life in a rescaling Europe.* Farnham: Ashgate.

Aradau, C., 2014. The promise of security: resilience, surprise and epistemic politics. *Resilience: International Policies, Practices and Discourses*, 2 (2), 73–87.

Bettini, G., 2017. Unsettling futures: climate change, migration, and the (ob)scene biopolitics of resilience. *In:* A. Baldwin and G. Bettini, eds. *Life adrift: climate change, migration, critique.* London: Rowman & Littlefield Publishers, 79–95.

Brand, F.S. and Jax, K., 2007. Focusing the meaning(s) of resilience: resilience as a descriptive concept and a boundary object. *Ecology and Society*, 12 (1), 23.

Bristow, G., 2010. Resilient regions: re-'place'ing regional competitiveness. *Cambridge Journal of Regions, Economy and Society*, 3 (1), 153–167.

Chandler, D., 2013. Editorial. *Resilience*, 1 (1), 1–2.

Chandler, D., 2019. Resilience and the end(s) of the politics of adaptation. *Resilience*, 7 (3), 304–313.

Christopherson, S., Michie, J., and Tyler, P., 2010. Regional resilience: theoretical and empirical perspectives. *Cambridge Journal of Regions, Economy and Society*, 3 (1), 3–10.

Davoudi, S., Brooks, E., and Mehmood, A., 2013. Evolutionary resilience and strategies for climate adaptation. *Planning Practice & Research*, 28 (3), 307–322.

Dunn Cavelty, M., Kaufmann, M., and Kristensen, K.S., 2015. Resilience and (in)security: practices, subjects, temporalities. *Security Dialogue*, 46 (1), 3–14.

Evans, B. and Reid, J., 2014. *Resilient life: the art of living dangerously.* Cambridge: Polity Press.

Ferdoush, M.A., 2022. Stateless' yet resilient: refusal, disruption and movement along the border of Bangladesh and India. *In:* D.J. Andersen and E-K. Prokkola, eds. *Borderlands resilience: transitions, adaptation, and resistance at borders.* London: Routledge, 106–118.

Frandsen, S.B., 2022. Schleswig. From a land-in-between to a national borderland. *In:* D.J. Andersen and E-K. Prokkola, eds. *Borderlands resilience: transitions, adaptation, and resistance at borders.* London: Routledge, 121–136.

Giuliani, G., 2021. *Monsters, catastrophes and the Anthropocene: a postcolonial critique.* London: Routledge.

Grove, K., 2014. Agency, affect, and the immunological politics of disaster resilience. *Environment and Planning D: Society and Space*, 32 (2), 240–256.

Grove, K. and Chandler, D., 2017. Introduction: resilience and the Anthropocene: the stakes of 'renaturalising' politics. *Resilience*, 5 (2), 79–91.

Hannonen, O., 2022. Mobility turbulences and second-home resilience across the Finnish-Russian border. *In:* D.J. Andersen and E-K. Prokkola, eds. *Borderlands resilience: transitions, adaptation, and resistance at borders.* London: Routledge, 90–105.

Healy, A. and Bristow, G., 2019. Borderlines: economic resilience on the European union's Eastern periphery. *In:* G. Rouet and G. Pascariu, eds. *Resilience and the EU's Eastern neighbourhood countries.* Cham: Palgrave Macmillan, 349–368.

Humbert, C. and Joseph, J., 2019. Introduction: the politics of resilience: problematising current approaches. *Resilience*, 7 (3), 215–223.

Iwabuchi, K., 2016. *Resilient borders and cultural diversity.* Lanham, MD: Lexington Books.

Koch, K., 2022. Cross-border resilience in higher education: Brexit and its impact on Irish – Northern Irish university cross-border cooperation. *In:* D.J. Andersen and E-K. Prokkola, eds. *Borderlands resilience: transitions, adaptation, and resistance at borders.* London: Routledge, 37–53.

Konrad, V., 2015. Toward a theory of borders in motion. *Journal of Borderlands Studies*, 30 (1), 1–17.

Konrad, V., Laine, J.P., Liikanen, I., Scott, J.W., and Widdis, R., 2019. The language of borders. *In:* S. Brunn and R. Kehrein, eds. *Handbook of the changing world language map*. Cham: Springer. DOI: 10.1007/978-3-319-73400-2_52-1.

Korhonen, J., Koskivaara, A., Makkonen, T., Yakusheva, N., and Malkamäki, A., 2021. Resilient cross-border regional innovation systems for sustainability? a systematic review of drivers and constraints. *Innovation: The European Journal of Social Science Research*, 34, 202–221. DOI: 10.1080/13511610.2020.1867518.

Kraemer, K., 2020. Longing for a national container. On the symbolic economy of Europe's new nationalism. *European Societies*, 22 (5), 529–554.

Laine, J.P., 2020a. Ambiguous bordering practices at the EU's edges. *In:* A. Bissonnette and É. Vallet, eds. *Borders and border walls: in-security, symbolism, vulnerabilities*. London: Routledge, 69–87.

Laine, J.P., 2020b. Safe European home – where did you go? on immigration, b/ordered self and the territorial home. *In:* J.P. Laine, I. Moyo, and C.C. Nshimbi, eds. *Expanding boundaries: borders, mobilities and the future of Europe-Africa relations*. London: Routledge, 216–236.

Laine, J.P., 2021. Beyond borders: towards the ethics of unbounded inclusiveness. *Journal of Borderlands Studies*. Epub ahead of print 10 May. DOI: 10.1080/08865655.2021.1924073.

Laine, J.P., Liikanen, I., and Scott, J.W., 2021. Changing dimensions of the Northern European security environment. *In:* J.P. Laine, I. Liikanen, and J.W. Scott, eds. *Remapping security on Europe's northern borders*. London: Routledge, 1–17.

Lamour, C. and Blanchemanche, P., 2022. A resilient *Bel Paese*? investigating an Italian diasporic translocality between France and Luxembourg. *In:* D.J. Andersen and E-K. Prokkola, eds. *Borderlands resilience: transitions, adaptation, and resistance at borders*. London: Routledge, 152–165.

Latour, B., 2018. *Down to earth politics in the new climatic regime* [ebook reader]. Cambridge, MA and Medford: Polity Press.

Lois, M., Cairo, H., and de las Heras, G., 2022. Politics of resilience . . . politics of borders? In-mobility, insecurity and Schengen "exceptional circumstances" in the time of COVID-19 at the Spanish-Portuguese border. *In:* D.J. Andersen and E-K. Prokkola, eds. *Borderlands resilience: transitions, adaptation, and resistance at borders*. London: Routledge, 54–70.

Mavelli, L., 2019. Resilience beyond neoliberalism? mystique of complexity, financial crises, and the reproduction of neoliberal life. *Resilience*, 7 (3), 224–239.

Paasi, A., Prokkola, E-K., Saarinen, J., and Zimmerbauer, K., eds., 2019. *Borderless worlds for whom? ethics, moralities and mobilities*. London: Routledge.

Prokkola, E-K., 2021. Borders and resilience: asylum seeker reception at the securitized Finnish-Swedish border. *Environment and Planning C: Politics and Space*. Epub ahead of print 14 April. DOI: 10.1177/23996544211000062.

Prokkola, E-K., 2022. Border security interventions and borderland resilience. *In:* D.J. Andersen and E-K. Prokkola, eds. *Borderlands resilience: transitions, adaptation, and resistance at borders*. London: Routledge, 21–36.

Ptak, T., Laine, J.P., Hu, H., Liu, Y., Konrad, V., and van der Velde, M., 2020. Understanding borders through dynamic processes: capturing relational motion from south-west China's radiation centre. *Territory, Politics, Governance*. Epub ahead of print 11 June. DOI: 10.1080/21622671.2020.1764861.

Ridanpää, J., 2022. Borderlands, minority language revitalization and resilience think-ing. *In:* D.J. Andersen and E-K. Prokkola, eds. *Borderlands resilience: transitions, adaptation, and resistance at borders.* London: Routledge, 137–151.

Rumford, C., 2012. Towards a multiperspectival study of borders. *Geopolitics,* 17 (4), 887–902.

Scott, J.W., 2020. Introduction to a research agenda for border studies. *In:* J.W. Scott, ed. *A research agenda for border studies.* Cheltenham: Edward Elgar, 3–24.

Star, S.L. and Griesemer, J.R., 1989. Institutional ecology, 'translations' and bound-ary objects: amateurs and professionals in Berkeley's museum of vertebrate zoology 1907–39. *Social Studies of Science,* 19 (3), 387–420.

Svensson, S. and Balogh, P., 2022. Resilience at Hungary's borders: between every-day adaptations and political resistance. *In:* D.J. Andersen and E-K. Prokkola, eds. *Borderlands resilience: transitions, adaptation, and resistance at borders.* London: Routledge, 73–89.

Wandji, G., 2019. Rethinking the time and space of resilience beyond the west: an example of the post-colonial border. *Resilience,* 7 (3), 288–303.

Wilson, T.M. and Donnan, H., 2012. *A companion to border studies.* Chichester: Wiley-Blackwell.

Index

Printed in the United States
by Baker & Taylor Publisher Services

Printed in the United States
by Baker & Taylor Publisher Services